Great Issues of the Day
Number Five
ISSN 0270-7497

THE COACHELLA VALLEY PRESERVE

The Struggle for a Desert Wetlands

by

Yvonne P. Tevis

Bibliography by M. Louise Reynnells

BORGO PRESS / WILDSIDE PRESS

www.wildsidepress.com

Library of Congress Cataloging-in-Publication Data

Tevis, Yvonne P. (Yvonne Pacheco), 1961-
    The Coachella Valley Preserve : the struggle for a desert wetlands / by
Yvonne P. Tevis.
      p. cm. — (Great issues of the day, ISSN 0270-7497 ; no. 5)
    Includes bibliographical references and index.
    ISBN 0-89370-332-X (cloth). — ISBN 0-89370-432-6 (pbk.)
    1. Coachella Valley Preserve (Calif.) 2. Uma inornata—California—
Coachella Valley. 3. Nature conservation—California—Coachella Valley.
4. Regional planning—California—Coachella Valley.   5. Reptiles—
California—Coachella Valley. I. Title. II. Series.
QH76.5.C2T48 1995   87-866
333.95'16'0979495—dc19   CIP

FIRST EDITION

CONTENTS

Illustrations
(Courtesy Jim Cornett, Stephen Willard and The Palms)

PREFACE

This is the true story of events in the Coachella Valley and all the characters are real. Although I have adhered to actual happenings and chronological order, my aim above all has been to tell the story of a local clash between the beliefs of conservationists and anti-conservationists (neither group right nor wrong), and I have not burdened the tale with many names or many details. A lot of people who worked hard both for and against the Coachella Valley fringe-toed lizard have not been mentioned.

I have followed instead the tale as told to me by the major characters. Each person colors his account with his beliefs. These beliefs stirred the clash in the first place, and these colorings enrich the story now.

All my information comes from newspaper and magazine articles, other published works, the stories of people I have interviewed, and my own observations about the Coachella Valley, having lived there most of my life.

—Yvonne Pacheco Tevis
San Francisco, California

LIST OF CHARACTERS

CHARLES ALLARD: Local physician and owner of the largest land parcel in the proposed Coachella Valley Preserve.

CAMERON BARROWS: Nature Conservancy's manager for the Coachella Valley Preserve.

BILL BONE: President of Sunrise Company.

PHILIP L. BOYD: Local conservationist who was instrumental in founding the Living Desert Reserve and the Philip L. Boyd Deep Canyon Research Center.

GEORGE E. BROWN, JR.:   California congressman who supported the federal listing of the Coachella Valley fringe-toed lizard.

LESTER CLEVELAND: Executive director of the Coachella Valley Association of Governments (C-VAG).

COACHELLA VALLEY FRINGE-TOED LIZARD: (*Uma inornata*)   The species of lizard listed endangered by the California Department of Fish & Game (CDFG), and listed threatened by the United States Fish and Wildlife Service (USFWS), and which caused great discord among the residents of the Coachella Valley.

LESLIE CONE: Indio area manager for the BLM.

JIM CORNETT: Curator of the natural sciences division of the Palm Springs Desert Museum.

JUSTIN DART: Former owner of 19,000 acres in and around the proposed Coachella Valley Preserve, which he sold to Charles Allard.

SID ENGLAND: Biologist and co-author of the CDFG's "Status of the Coachella Valley Fringe-toed Lizard" who took aerial photographs of the Coachella Valley to measure the disappearance of habitat.

GERALD (JERRY) FORD: 38th president of the United States, a resident of the Coachella Valley and a supporter of the Coachella Valley Preserve.

GAIL KOBETICH: Director of the Sacramento Endangered Species Office of the USFWS.

PATRICIA "CORKY" LARSON: The Coachella Valley's representative to the Riverside County Board of Supervisors.

JERRY LEWIS: California congressman active in supporting flood control.

WILBUR MAYHEW: Biology professor at the University of California, Riverside who spearheaded the battle to place the Coachella Valley fringe-toed lizard on the endangered species lists.

AL McCANDLESS: California congressman who introduced to the U.S. House of Representatives appropriations for the Coachella Valley Preserve from the Federal Land and Water Conservation Fund.

STEVE McCORMICK: Field representative for California Nature Conservancy.

PEARL McCALLUM McMANUS: Daughter of the founding family of Palm Springs.

AL MUTH: Manager of the Coachella Valley's Philip L. Boyd Deep Canyon Research Center, director of the Coachella Valley Ecological Reserve Foundation (C-VERF) and watchdog in the valley for the fringe-toed lizard.

DAVID PACKARD: Member of the Board of California Nature Conservancy.

PAUL SELZER: Attorney for Sunrise Company who was greatly influential in coordinating the objectives of both developers and conservationists.

EDWARD WASSERMAN: President of the Thousand Palms Chamber of Commerce.

DON WEAVER: Design Engineer for the San Bernardino office of the California Department of Transportation who helped locate a site with sufficient blowsand for a preserve for the Coachella Valley fringe-toed lizard.

LOWELL WEEKS: General manager and chief engineer of the Coachella Valley Water District.

GRANT WERSCHKULL: Nature Conservancy's local fundraiser and director for the Coachella Valley Preserve.

GEORGE WHITNEY: Desert realtor who represented Dr. Charles Allard.

PAUL WILHELM: Writer who lives at Thousand Palms Oasis and son of Louis Wilhelm, one of the original settlers of the Coachella Valley.

RIC WILHELM: Land manager for the Coachella Valley Preserve and grandnephew of Louis Wilhelm.

JAN ZABRINSKY: Manager of the Philip L. Boyd Deep Canyon Research Center and director of C-VERF before Al Muth took over both positions.

GLOSSARY

balance of terror: A respect Coachella Valley developers and conservationists held for each other's capabilities of retaliation, and which compelled both parties to keep their sides of the bargain and create a desert preserve.

blowsand: Fine, loose, wind-borne sand which is the essential feature of the Coachella Valley fringe-toed lizard's habitat.

Bureau of Land Management: (BLM) The agency of the federal Department of the Interior which has jurisdiction over certain public lands.

California Department of Fish and Game: (CDFG) The state department which manages wildlife.

California Department of Transportation: (Caltrans) The state department which maintains state roads. It played a key role in determining an area which continually receives blowsand and which would thus make a good fringe-toed lizard preserve.

California Nature Conservancy: The nearly autonomous California chapter of Nature Conservancy.

Coachella Valley: A desert valley in Southern California, a fashionable winter resort, and the scene of a dramatic struggle to preserve wilderness.

Coachella Valley Association of Governments: (C-VAG) An organization of representatives from some Coachella Valley city councils plus the Riverside County Board of Supervisors which discusses valley-wide issues.

Coachella Valley Ecological Reserve Foundation: (C-VERF) A group organized by Dr. Wilbur Mayhew to work for a desert preserve.

Coachella Valley Habitat Conservation Plan: The name of the specific conservation plan for the Coachella Valley to protect the Coachella Valley fringe-toed lizard.

conservation plan: A plan for saving an endangered species. An applicant for an ESA 10(a) permit to incidentally take a federally protected species prepares a conservation plan which ensures that the taking of a few individuals will be outweighed by mitigation measures prearranged in the plan.

critical habitat: The area which the federal government finds essential to the preservation of a federally protected species. The critical habitat does not necessarily comprise the entire area occupied by the species. Also, it can include areas outside the species' habitat if they are necessary for conserving the rest of the critical habitat.

Deep Canyon Transect: A study area used by Deep Canyon Research Center researchers which embraces the Deep Canyon drainage and adjacent desert and mountain habitats.

Desert Protective Council: An organization founded in 1955 which works for the protection of California, Arizona and Nevada deserts.

Desert Riders: A group of Coachella Valley horseback riders who regularly get together for organized rides through local desert regions.

economic impact report: Before a species is considered for federal protection, USFWS staff must prepare a study analyzing the economic effect which federal protection of a species may have on the human population in and around the species' habitat.

endangered: A term describing a species placed on state or federal endangered species lists. California defines a species as endangered if the species is in immediate peril of extinction. The USFWS defines a species as endangered if it is in peril on parts or all of its habitat. Endangered signals the highest level of protection for a species on both the California and federal levels. The Coachella Valley fringe-toed lizard was declared endangered by California on June 27, 1980.

Endangered Species Act: (ESA) A federal law established in 1973 to provide a means of protecting national wildlife by naming certain species threatened or endangered and by then requiring the USFWS to develop a recovery plan for these species. The act was amended in 1978 and 1982.

Endangered Species Act 10(a) permit: (ESA 10(a) permit) If a species is federally protected, which means no one may harm it, and yet someone wishes to conduct an activity such as research or development which will in fact harm the species, he may apply for this permit. To be eligible for the permit he must fulfill certain conditions as set up in section 10(a) of the ESA which aim to ensure the survival of the species.

endangered species list: A state or federal list of protected species.

environmental impact report: (EIR) Cities and counties require a developer to study the potential effects that a planned development may inflict upon the environment around the proposed site. Included in this report is any request to develop.

Federal Register: A periodically published report announcing federal proposals and actions. Any proposal to list a species on a federal endangered species list must, under the ESA, be published in the Federal Register.

Friends of the Fringe-toed Lizard: A group of Coachella Valley realtors, lawyers, and businessmen bent on preventing the Coachella Valley fringe-toed lizard from becoming a state-protected species.

historic range of the Coachella Valley fringe-toed lizard: The habitat which the lizard probably had occupied before human development reduced that habitat.

incidental take: When an applicant for an ESA 10(a) permit receives the permit, he then is allowed to incidentally take a protected species. In other words, harassment of the species is not his goal, but a consequence of his main activity, such as research or development. The harassment is merely incidental. (*See also* **take**.)

Land and Water Conservation Fund: A federal fund from which Congress appropriates money for acquisition of wildlife habitat.

listing: Used as a noun, it is the act of a species being added to state or federal endangered species lists.

Living Desert Reserve: A 1,200-acre reserve in Palm Desert established to preserve natural desert and to encourage an interest in the desert among Coachella Valley residents and visitors.

Lizard Club: The unlikely coalition in the Coachella Valley of developers, conservationists, and members of local, state, and federal agencies which worked out a means to save the Coachella Valley fringe-toed lizard, and to preserve natural desert.

McCallum Desert Foundation: A foundation established in the will of Pearl McCallum McManus endowing grants to local organizations.

mitigation: Measures taken by the recipient of an ESA 10(a) permit to make up for incidental taking of a federally protected species.

mitigation fees: In the Coachella Valley, developers pay a fee per acre if they develop part of the Coachella Valley fringe-toed lizard's critical habitat. The money goes to buy a preserve in another section of the critical habitat.

Nature Conservancy: A private, national, conservation organization which buys private preserves throughout the United States rather than lobbying for governmental protection of wilderness.

Palm Springs: A resort city in the Coachella Valley. Used in this book, however, to describe the whole resort area of the Coachella Valley, including the communities of Cathedral City, Indian Wells, Indio, La Quinta, Palm Desert, Palm Springs, and Rancho Mirage.

Palm Valley Country Club: The Sunrise Company development which was temporarily halted by amendments to the ESA in 1982. Sunrise Company later became active in working out a solution to satisfy conservationists and developers.

Philip L. Boyd Deep Canyon Research Center: A desert reserve and research center in the Coachella Valley administered by the University of California, Riverside.

protected: Describes a species declared rare or endangered by California, or declared threatened or endangered by the federal government.

R. A. Glass Company: A company which violated the ESA and planted grapes on the critical habitat of the Coachella Valley fringe-toed lizard.

rare: The second highest of two levels of protection awarded a species by the state of California. A species is rare if its numbers have so dwindled that changing its habitat might jeopardize its survival.

recovery plan: When a species is placed on the federal endangered species list, the USFWS, under the ESA, must develop this plan to conserve the species and prevent its extinction.

Richard King Mellon Foundation: A foundation which, out of its $25-million wetlands conservation challenge fund, granted California Nature Conservancy $2 million to buy Thousand Palms Oasis, a desert wetlands located in the Coachella Valley.

Sunrise Company: The development company which has built many country clubs in the Coachella Valley, and which took on the task of organizing a desert preserve.

take: The word used in the ESA to describe actions which adversely affect the survival of a species. The ESA defines to take as to "harass, harm, pursue, hunt, shoot, wound, kill, trap, capture, or collect, or to attempt to engage in any such conduct."   (*See also* **incidental take**.)

threatened: The second highest of two levels of protection awarded a species by the USFWS. A threatened species may be in jeopardy in the kind of future man foresees. The Coachella Valley fringe-toed lizard was declared threatened on September 19, 1980.

United States Fish and Wildlife Service: (USFWS) The federal agency of the Department of the Interior which manages wildlife on both private and public lands.

I.

A HANKERING FOR OPEN SPACE

Because of a $25 million deal to allow development to take over a desert valley, a California desert lizard will continue to dive beneath the hot surface of sand. Windstorms will whirl grit, pile dunes, smooth swales. Roadrunners will scamper over the dunes, a trail of Xs crisscrossing the sand waves. Coyotes will trace the scent of game, smell water from an oasis and go to drink where water wells to the surface through cracks in the San Andreas earthquake fault, where palms with thick skirts of dry fronds cluster around the water and rustle and creak in the wind. The green tops spin.

All this is part of the Coachella Valley, the desert most people know as Palm Springs, California, a sunny, winter golf resort for Easterners come shivering from November snows. Magazine ads for the Coachella Valley picture emerald golf courses, lakes, condominiums, and richly dressed people who pose in the foreground of a sunset capping blue mountains. An enchanting night beckons....Publicity brochures rarely tantalize potential visitors with the rough and harsh desert.

An incredible concentration of wealth has settled in the desert cities of the upper Coachella Valley—Palm Springs, Cathedral City, Rancho Mirage, Palm Desert, Indian Wells, and La Quinta. (See Map I) Flat and easily developed, land in potential resort areas yields high prices. Therefore, if one wants to develop, land set aside as a preserve is dollars tossed in the dump. Since as far back as the end of World War II, the resort's ever-growing development has frightened people who admire the untouched desert, and who have dreamed of a large, desert preserve.

17

The local battle over how desert land should be used—to be developed or preserved—exploded in the 1970s, a biology professor leading the conservationists' front. The rancorous clash divided residents. In the remarkable end to this tale, developers and conservationists actually work together to save the natural desert, a finale especially remarkable in a wealthy resort where growth brings tourism, and more wealth.

By telling a story of a dream envisioned, of a quest of ups and downs, and of character clashes, this book means to illumine the strategy which Coachella Valley conservationists followed to preserve part of their desert wilderness. Listening to a tale, and understanding why and with what tools people oppose conservation, will help environmentalists execute another successful plan to save another wilderness.

In this story, a lizard—the Coachella Valley fringe-toed lizard, *Uma inornata*—is the pivotal character. Conservationists used the lizard, whose rapid extinction they feared, to obtain governmental protection for the lizard's habitat, expecting thereby to create a desert preserve. The conservationists proposed that the California Department of Fish and Game (CDFG), a state agency, put the Coachella Valley fringe-toed lizard on its "rare" or "endangered" species lists, and that the United States Fish and Wildlife Service (USFWS), a federal agency, put the lizard on its "threatened" or "endangered"species list. The agencies must first announce the proposal to list the species, then conduct an environmental survey to analyze whether the species needs protection, and later hold public meetings airing opinions for or against the listing. If the agency lists the species, public funds can be used to buy habitat preserves.

Also, if the agency lists the species, environmental laws restrict what people can do both to the species and also within its habitat. Realtors, developers, and property owners opposed a listing of the Coachella Valley fringe-toed lizard because they feared that governmental protection of the lizard's habitat would freeze urban growth, and hence, valley prosperity. Local people split bitterly over ideology:

to conserve the raw desert, or to develop a resort; to admire the adaptations of animals, or man's comforts; to perceive the desert as rich with wildlife, or as a wasteland; to encourage federal, or local decisions over land use; to promote a reptile, or human beings. They split over the question "Why save wildlife?"

In the Palm Springs area many people thought listing the lizard would mean a stricture on development and human progress itself. Traditionally, Americans have applauded progress west—the development of railroads, mines, and mills, power lines, dams, six-lane freeways, cities, and country clubs. A biologist in Palm Springs suggested this view of progress stems from a Judaic-Christian tradition: "Be fruitful, and multiply, and replenish the earth, and subdue it: and have dominion over the fish of the sea, and over the fowl of the air, and over every living thing that moveth upon the earth."[1] He added, though, that the Old Testament also calls for good stewardship of the earth. In any case, Americans have heard often enough that they possess a lust to delve into, and cut open frontiers, and to build a progressive nation.

Yet Americans cling to another vision of America—a land of open spaces. The new world seduced explorers with the promise of horizons untouched by man's finger, a place where the individual could conduct his life by his own laws. A man could be a man. That endless, unmarred, challenging frontier has disappeared. The nationwide lungful of exhaust emitted from city buses coast to coast proves that. Now, at the end of the twentieth century, the United States could be as crowded as any European country except for the old American thirst for space which prompts national protection of wilderness on a grand scale, making preservation of wilderness an American phenomenon.

The American promise of space no longer signifies the huge and unexplored, but instead, a confined wilderness for an individual himself to discover—a swamp, an island, a forest, a desert, a wilderness preserved, space where the individual stands out from a seething populace, where twentieth-century Americans touch bedrock, that which no other human mind created (unlike skyscraper horizons). The promise of space we've sometimes kept is to preserve for all Americans chunks of our wild places.

Grand words about the needs of the human soul don't in themselves win conservation wars, and didn't in the Coachella Valley. The innate pride of most Americans in their national, natural treasures lies like a seed which can, under the right conditions, germinate into dollars and votes. To make the seed grow necessitates a practical strategy. With one wilderness secured, others vanish. Conservationists working on one project snap that file shut and rush on to the next. Yet reflections of past success sharpens the memory and triggers new ideas.

A story like that of the Coachella Valley Preserve reveals strategy behind one fight to save wilderness, and an example for the next. If developers, largely the Sunrise Company, hadn't finally joined the conservationists, the flat, sandy area of the preserve would now be vineyards and housing tracts, while camping trailers, picnic tables and toilet facilities would crowd beside the desert oases at the heart of the preserve. (See Map III) The amazing compromise between developers and conservationists reached national news. On July 9, 1984, the *Wall Street Journal*[2] told the story on the front page: to develop their land, developers for thirty years will pay mitigation fees which go towards buying the preserve. In exchange, the developers will be free to build upon the remaining open land, whether it is the habitat of the protected lizard, or not. Strategy and fight oiled this democratic wheel of compromise which for decades hadn't budged.

Nowadays, when people think of Palm Springs, they picture drinks by the pool, shopping, golfing, tennis, sunshine—the general good life. Some sixty country clubs, with more being developed, sprawl across the desert. Nearly all brag acres of golf courses and lakes. Waterfalls at the entrance gates lure buyers because water in the desert seduces like magic. At one club, houses surround a lake; private docks extend from each house for sailboats.

Yet in the old days, long ago, the resort lifestyle centered around dude ranches. Hollywood movie stars came to the desert to ride horses across the sands, and sweep their gaze across the empty valley to the brown hills, and the blue ranges beyond. How do you combat modern Palm Springs where people want golf courses with lakes, not sand? There, in that valley, the answer: fight water with water. People like water. It refreshes; it's recreational. Oases in the desert exude a romance.

In this story, California Nature Conservancy receives a $2 million rollover loan from the Richard King Mellon Foundation's $25 million wetlands preservation project to buy a desert oasis. Nature Conservancy utilizes the magical appeal of water in the desert and takes charge of creating a huge, $25-million desert preserve with the oasis of palms and natural pools at its center. The public reaction to a proposed preserve swings from viewing it as a wasteland crawling with pesky reptiles, to a pretty, natural park of which all the burgeoning desert cities may boast.

The story about creating the Coachella Valley Preserve does not offer the perfect guide to save wild lands. Once, the fringe-toed lizard slid over 267 square miles of sand. Now, only 127 square miles of lizard habitat remain, one half of which development is currently destroying.[3] The Coachella Valley Preserve itself will comprise roughly twenty square miles, much of it not lizard habitat.[4] (See Map II) Who has compromised more? Development will encounter few problems as it swallows the rest of the valley.

Nevertheless, the preserve will signal a home saved for desert wildlife, and a reaffirmation of one principle so essential to democracy we often forget it: variance among lifestyles must be allowed. The story shows a conservationist that while he compromises with developers he must also take the offensive, do the unexpected, and sometimes be as outlandish as to find a wetlands in the desert.

II.

THE BULLDOZED DESERT

An aura of function permeates Dr. Wilbur Mayhew's office at the University of California, Riverside (UCR).   Biology professors proverbially bedeck their cramped university quarters with skeletons, stuffed birds, and rock specimens.   Dr. Mayhew hasn't bothered.   Several tall, metal bookshelves fill the center like the reference stacks in a library—lots of sets and dark green bindings. On either side sits a big desk surrounded by ominous, grey, filing cabinets which match the grey floor.   Hung high on the walls in an exact row are black and white pictures of birds.   Below them hangs one frame of a lizard.

Biology professor Mayhew, who once thought no place on earth could be more ghastly than a desert—yet battled thirty years to protect one—and who specializes in birds—yet battled thirty years to save a lizard—talked nonstop.

An ebullient, ever-active person, he jumped up in the middle of his story and grabbed a well-worn flyswatter. "I hate flies. I grew up on a dairy farm and can't tolerate flies." He whacked the swatter through the air.   "If it'd just land somewhere other than on you." He looked frustrated.   "If I could get it against the light, sometimes you see better and can hit them that way."

Other winged creatures, especially birds, delight him, and he likes to slosh through marshes and wetlands. While at UC Berkeley he studied how moisture affects reproduction among mallard ducks. A large photograph of a mallard hangs on his wall.

He said he suffered six months during World War II in the Sahara and vowed to never go to another desert again.   "It was awful," he said.   "Once, we had to tramp alongside a river for eighteen miles. Eighteen miles and I counted only three plants and four birds. It was an awful place."

But his first year at UCR found him going back to a desert. In 1955, as the first professor to teach the university's new field biology class, he had to find some field easy to get to where the animals and plants hadn't been disturbed. By car, the Coachella Valley lies an hour-and-a-half from Riverside. In 1955, the desert seemed to extend for mile after unbroken mile.

Dr. Mayhew figured that since he was in the desert he might as well study how moisture affects the reproduction of desert birds, and so, with his classes, he set out upon exploratory hikes across the sand.

At Berkeley, his method of research had entailed capturing pregnant females in the wilds, transporting them to his laboratory and running tests on them there. Dr. Mayhew confidently began his desert search. He encountered a problem. He'd find one nest and then walk miles before he'd come across another, much less a pregnant female waiting. Desert dunes don't teem with birds like the wetlands of Northern California do.

"What I saw instead were lizards—everywhere." Lizards popped out from the sand, scuttled across the tops of dunes, scrambled under wind-twisted mesquite trees. A biologist watches life—anywhere—in a marsh, in a desert. "I figured I could apply my same question to lizards, how moisture affects their reproduction."

Yet another problem baffled the wetlands biologist displaced in a desert.

"Field books about birds describe precisely the birds' nesting and egg incubation periods, what time of the year females lay eggs, and when to trap them for laboratory studies," he said. "I went through lizard book after lizard book. They didn't even tell whether a species lays eggs or births live. They just weren't geared for this type of study." He saw he would have to gather all the initial information himself before he could commence his moisture studies.

Taking his classes, he spent a year combing the dunes, compiling notes on lizards of the Coachella Valley. The desert worked its spell upon him. He marvelled how desert creatures carry out life superbly in that harsh clime. Among his notes, he collected vast information on a master of desert living—the Coachella Valley fringe-toed lizard.

A second year rolled along as Dr. Mayhew re-checked his findings. The new notes contradicted all his research of the first year. "As you well know, every year in the desert the weather changes from the year before. I spent five years checking data." In his thirtieth year to escort biology classes to the desert, he said he has not noticed a constant pattern in desert weather.

In 1958, three years after his arrival in the desert, the National Science Foundation awarded Dr. Mayhew a grant to study the effects of climatic changes upon reptiles.

"That's about when I began to lose research sites to the dune buggies and bulldozers." Back then, few people believed the expanse of desert would shrink, because the sense of space intoxicates. Dr. Mayhew couldn't comprehend people would really want to live across the whole desert floor.

"At first, as bulldozers gobbled up one spot, I'd just move my classes to the next dune over. But then it got so that repeatedly we'd go back a week later and find *that* site all cleared for development and we'd have to move again."

Development and dune buggies shunted around other UCR professors. Irv Newell, tracing the lifecycle of a species and studying the animals of a specific research site as his models, one weekend popped down to Palm Springs to check his site. Bulldozers had wiped it out. He had to start the long research all over so his study would be contiguous.

Jim Cornett, the natural sciences curator at the Palm Springs Desert Museum, upon hearing Dr. Mayhew's story said, "UCR professors aren't the only ones who lost sites." In 1977 he had barely completed his master's degree before his research spot became a country club.

"I'd finished, but I'd wanted to continue my research a few years more. Development took care of that."

25

Indians and desert creatures had once lived alone in the foothills and canyons and oases of the Coachella Valley. In the late 1800s, white settlers moved to the valley they named Palm Springs.[5] Ranchers planted dates, citrus, and vineyards, and rows of tamarisk trees to halt the wind from dumping sand in their fields. In the 1940s Hollywood movie stars descended upon the desert. Dude ranches became the rage. Movie directors filmed westerns in the sand and by oasis pools. The glamorous life lured more people. Cities popped up like spots on a toad's back all the length of the valley.

By the time Dr. Mayhew was dune-hopping, golf courses, swimming pools, condominiums and eternal sunshine signified the new Palm Springs, the playground for stars such as Bob Hope, Frank Sinatra, and the Marx Brothers, for retired politicians like Gerald Ford, Spiro Agnew, and for ambassadors, Texas oil magnates, and golf and tennis pros. Yellow bumper stickers, tee shirts and tote bags all featuring a black sun wearing dark glasses and reading *P.S. I LOVE YOU.*

As the hordes rush in to enjoy an easygoing lifestyle, the tranquility slips out, the sigh of air behind a closing door. A stop light bottles up traffic every mile along the main highway which runs the length of the valley and some citizens proclaim, "Every additional stoplight is a step backwards." Weekenders grumble in the express lines of markets: "Guess everyone else decided to come out when I did." They pay for their vodka and peanuts and hurry to their cars in the parking lot where everyone stop-and-start drives, vying desperately to get in front of the others and onto that highway....

The most packed times come during the five days of the Bob Hope Desert Golf Classic each winter and during Easter week, when to drive one block in Palm Springs takes forty-five minutes. Residents swear that next year they won't step foot outside their homes.

As more people move to the desert, or buy second homes, the walls of the condominiums spread. Each country club acts like its own feudal city complete with security guards, golf courses, lakes, tennis courts, and a clubhouse with a dining room. Coming home for vacations, date groves spirited away, sand dunes carted off, roads widened into thoroughfares, roads blocked for construction, roads re-routed, country clubs and shopping centers sprouting, I discovered a strange new valley every time.

In the old days, people used to flee for the summer, the streets crackling emptily in the heat. Not anymore. The hotter, the better—in a pool, or a restaurant. As the cities swell, the space and solitude, and barrenness, the essence of desert for the people who love it, withers.

Dr. Mayhew got mad. He didn't have to buy groceries in Palm Springs. He didn't go too much into town. He was mad because he'd lost fifteen sandy research sites to development which transformed the desert, and to dune buggies that mutilated it.

For desert research, UCR uses the Deep Canyon Transect. If you stretch an imaginary road down from the mountains on one side of the valley, across the sandy floor, up the mountains of the other side, the strip includes every habitat in the desert. Thus, to continue to study the whole desert environment, an example has to be preserved of each habitat of the desert mountains and the valley, although not in a contiguous ribbon. The university relied upon federal Bureau of Land Management (BLM) and state lands in the mountains, and upon its own Philip L. Boyd Deep Canyon Research Center in the canyons. The university lacked protected sand habitats. Dr. Mayhew said destruction of sandy sites crippled research. Before, one biologist would make his studies and write a book for the university press. Then another biologist could, even years later, pick up where the first had left off, and go back to the same site. Now, continuity among university authors would end unless researchers could save a segment of sand habitat for their work. Dr. Mayhew began thinking "preserve."

The preserve could exist only if it were guaranteed sand and more sand and more sand. In the Coachella Valley, floods carry sand from canyons to alluvial fans. Winds, which generally blow from the west down to the east, drive the sand from the San Gorgonio mountain pass and the alluvial fans across the desert, thus replenishing the sand fields.[6] Any agriculture or development between the alluvial fan of a canyon and the preserve would slow the wind, pin down sand, and block sand from reaching the preserve. After awhile, the sand of the preserve would drift southeast down the valley to the Salton Sea, leaving a hard, truly inhospitable wasteland behind.

Dr. Mayhew wanted some area that would receive a constant influx of sand, either from wind or floods. Because it combats desert weather on the highways, the California Department of Transportation (Caltrans) has had to study the natural movement of sand, wind and water. Working closely with Don Weaver, a design engineer for Caltrans, Dr. Mayhew narrowed his preserve location to two sites on the northern side of the valley, away from the popular areas of development, each location backed by a canyon through which floodwaters would bring sand. That way, the sand-loving creatures of the sand habitat wouldn't have to rely solely upon the wind to provide sand when, after all, someday the whole pass area could be covered with fields and homes and wind could blow without raising sand storms.

Dr. Mayhew disqualified one idea, a Whitewater Canyon preserve. Whitewater Canyon had become a political cause, people demanding dams for flood control.

"Thousand Palms Canyon wasn't threatened yet with damming," said the biologist, "and the area is even farther away than Whitewater from the cities, and so less disturbed, and cheaper.

"Considering all this, and that Caltrans backed me up with proof that there would be a constant sand source feeding the area, that's where the final decision took us, around 1972." (See Map I) Dr. Mayhew marked his area on a map—Thousand Palms Canyon, replete with dunes and silt flats and arroyos. He was ready to buy.

The university had a program for establishing reserves, but it didn't have much money. "It was still just getting onto its feet," said Dr. Mayhew. More than fifteen years had passed since he first came to the desert and the burgeoning communities looked nothing like they had then.

With no time to wait for the university to find funds, and with the land being under county jurisdiction, he took his proposal for a preserve to the county seat in Riverside.

"I was naive," the biologist said. "The situation seemed clear, the need was obvious. I'd done all the work and had pinpointed a spot. The planning department could zone the area as an ecological preserve."

He assumed the county would agree easily that a natural park, maintained by the winds, could only enhance the prestige of the county.

"The planning director was new. I didn't know him, and he didn't want anything to do with a preserve. Building was too profitable then.

"But I had friends in the California Department of Fish and Game. I asked them for ideas. They said, unofficially, not as department spokesmen, 'If the county won't talk ecological preserve, find some species in your preserve and get it on an endangered species list.'"

So Dr. Mayhew began a new hunt.

III.

THE COACHELLA VALLEY FRINGE-TOED LIZARD

The white dunes heave and fall across the Coachella Valley. A lizard scuttles on top of a sparse dune, and in a flash, disappears.

As Dr. Mayhew explored the dunes in the early 1950s, a distinctive lizard astonished him with its vanishing act. His fascination grew, and his research centered more and more on this lizard. July and August days crackled at 120 degrees. While human beings dove into swimming pools, Dr. Mayhew watched the little Houdini—the Coachella Valley fringe-toed lizard, *Uma inornata*, dive below the surface of the sand in the shadow of a mesquite bush, and swim.

Often, people who don't know much about the desert think of it as a hell in which creatures can barely survive, and that while wildlife would prefer to move to the mountains, it adapts to the desert as best it can, and suffers. The obvious response is that for desert wildlife the most perfect spot to live is the desert. The fringe-toed lizard cannot survive in the mountains or golf courses or date groves, but dwells in the soft blowsand of the desert sand fields.

Over the millennia, this sand-loving lizard has gathered together a remarkable set of physical traits to perfect its desert existence. "Other animals have these same morphological adaptations," said Jim Cornett, the natural sciences curator at the Palm Springs Desert Museum, and who has published articles about his own studies of the lizard, "but not in North America, not in one package like the fringe-toed lizard."

31

So in the Coachella Valley, this remarkable, handily-packaged lizard employs every part of its body in the streamlined sand dive. With a head shaped like a pointed spade, the lizard takes a step backwards, and plunges into and under the sand, headfirst.[7]

Two eyelids shield each eye, the bottom lid being transparant so that the lizard perceives fluctuations in light and shadow in the world above.[8]

The pressure of the sand holds a loose flap of skin against each ear. Even so, the lizard can hear insects on the surface.[9]

To seal the mouth, the upper jaw bites down over the lower jaw.[10]

The nostrils halfway close to keep from getting clogged with sand. The nasal passage itself is U-shaped, like pipes in the kitchen sink. Any persistent grains of sand are trapped in the curve of the U. When the lizard emerges from the ground it snorts, a blast of air shooting out the trapped sand.[11]

Beneath the ground, the lizard breathes the air around the grains of sand.[12] The lizard stays underground only a short while, a couple of inches deep, and its swim goes only as far as a few inches. Far enough, however, so that when the lizard senses the presence of a human being ("I really can't say exactly what happens when other predators approach," Jim Cornett laughed, "because I can't be a coyote."), or coyote, roadrunner, or a shrike armed with a lethal, hooked beak, the lizard can dive, swim a few inches, and the hunter can't tell where the lizard has hidden. Sometimes, if you poke your fingers in the sand around where the lizard vanished, it will pop out and race off at great speed towards a mesquite bush shelter.

So that the lizard can glide quickly and smoothly both under and over sand, it has silky skin of small, rounded scales which cuts down on resistance.[13] Larger scales fringe each toe—hence, fringe-toed lizard (although Cornett prefers the other name, Coachella Valley sand lizard). These scales hang loosely to the toe to provide traction as they splay out when the lizard pushes backwards, and tuck in when it scurries straight ahead.[14]

Beneath the sand, the lizard rests safe from coyotes, meanwhile lowering its body temperature. On very hot days, said Cornett, the underground temperature of the sand isn't cool enough to bring down the lizard's body temperature, so it seeks shade, often burrowing in the sand, or eating the insects clinging about the bush.

The lizard has adapted both to an underworld, and the realm above ground where it relies on a different disappearing trick—camouflage. It's skin, white with speckles of black, blends exactly with the sand. In the upper world, a male lizard stakes out a small, sandy area for its own territory—a hump for a vantage point—and attacks any trespassing lizards.[15] The lizard resides a king of the blowsand habitat, perfectly suited to the desert and perfectly happy.

However, the fringe-toed lizard's habitat does not embrace the whole Coachella Valley, but only those parts characterized by loose, fine blowsand such as the mesquite-crowned dunes and the creosote-spotted sand flats. Coarse sand makes it hard for a lizard to push below the surface and speed along the top. Also, the compacted, clay ground of the valley where floodwaters have dumped silt, renders sand diving impossible.[16] The fringe-toed lizard needs blowsand.

The loose sand would flatten and harden if wind and water didn't continually replenish the habitat with new, soft sand. Floods carry crumbling, crushed rock from the foothills into the valley, and windstorms lift the finer sand into the air. The sand falls against creosote bushes, forming small hillocks. Mesquite dunes grow as the sand collects right up to the top of these small trees.

Daily, humans destroy the desert. Bulldozers remove dunes, leveling ground for farm fields and buildings. Dune buggies roar up, over, and down dune after dune. Although wind and water may eventually restore those dunes gouged by dune buggies, the rebirth of habitat takes a long time—longer than a human being lives. Meanwhile, the lizard's kingdom vanishes, and the rapid growth of a resort stymies the ecological revitalization of the desert.

Desert also disappears as people attempt to halt the wind itself—wind that stings eyes with sand, muddies swimming pools, pits windshields, makes roads slippery, cuts down on drivers' vision. The highway department plants windbreaks of salty tamarisk trees. Huge hedges encircle ranches. Many residents applaud the transformation of wilderness into squash fields or shopping centers simply because plants and pavement trap sand.

Agriculture and urban development aren't the only culprits. Two rapacious foreigners, an Asiatic tumbleweed and a Mediterranean tamarisk, have entrenched themselves in the Coachella Valley.[17] (Ironically, cartoonists draw the tumbleweed, or Russian thistle, as a symbol of the howling and wind-swept desert of the American West.) The immigrants have spread in thick mats, choking the movement of sand. Tumbleweeds, windbreaks, dune buggies, and condominiums don't discern between fringe-toed lizard areas and other habitats. They transfigure all.

The loss of sand habitat dooms many animals found only in the Coachella Valley such as the horned toad, the red velvet mite, the desert cockroach and the fringe-toed lizard, along with plants such as the Coachella Valley milk vetch, or the flat-seeded spurge. As they began choosing a species they could have declared endangered, Dr. Mayhew and his friends sought a politically appealing animal. Laymen are more likely to sympathize with the fate of an animal than with some dusty shrub, or bug. Cockroaches rarely get any attention other than cries of "Filth!" and squirts of roach poison. Even though Halloween movies defame lizards, snakes and other creepy-crawlies, children keep pet lizards. A lizard seems to have a cocky stance and a cheerful attitude. People think it's cute.

Besides emphasizing the charming qualities of a species to rouse popular support for it, another technique is to publicize its usefulness for man. The Coachella Valley fringe-toed lizard also boasted this qualification as it had aided research in mental instability. The fringe-toed lizard, as do many other lizards, possesses three eyes. In the late 1940s, Dr. R. C. Stebbins studied the third eye (an organ on the top of the head which perceives light and darkness) of the Coachella Valley fringe-toed lizard for his Ph.D. Later, his curiosity unsated, he and another biologist, Dr.

Richard Eakin, trapped Western fence lizards, removed their third eyes and released them into the habitat again. The researchers found that the lizards' time cycle went out of whack, perhaps explaining why they seemed weaker and succumbed to disease more readily than before. The reproductive cycle also revolved more quickly.[18]

The two scientists hypothesized that the third eye sends light to the pineal gland, an organ which regulates behavior. Nearly all vertebrates possess pineal glands. However, the higher life forms have evolved means other than a third eye of getting light to the pineal gland, and only a few tadpoles and frogs, and several hundred species of lizards have kept the third eye, which almost all prehistoric animals possessed.[19]

Julius Axelrod won a Nobel Prize in 1970 for his research on the effect of the pineal gland upon the behavior of rodents. Crediting Stebbins and Eakins' third eye research as a major source of information for his experiments, he discovered that the pineal gland and pituitary gland work together to balance the stimulation or repression of sexual behavior. Because of the third eye research which included the study of the Coachella Valley fringe-toed lizard, researchers are now exploring the possibility that a malfunction in the pineal gland could trigger mental disorders such as manic-depression and schizophrenia.[20]

For Dr. Mayhew, an animal doesn't have to be cute or useful to deserve protection. Life itself fascinates Dr. Mayhew, how it moves, how it lives...and Jim Cornett agreed.

"The Endangered Species Act," Cornett said of the federal law protecting endangered species, "of course means an enlisted species warrants a high priority for habitat acquisition. But I want to save *all* life—because it's there. Development can go on a hundred years from now, but not the environment. I'm not saying you can't develop at all, just not in specific spots. We can't recklessly doom options still open to us."

The specific spot for the Coachella Valley fringe-toed lizard means the Coachella Valley, the only place in the world where it exists. "If people won't save what they have the whole world population of, what *will* they?" Dr. Mayhew asked. Said Cornett, "I can't believe an intelligent person would eradicate a species when you can't show direct harm to people *because* of protecting it."

But some think "harm" means denying people condominiums, tennis courts, golf courses, and shops—a resort where they can relax and enjoy the rewards of hard work.

Dr. Mayhew gathered his resources to combat such thinking. Through combining his previous research with that of other biologists, he possessed more information on the Coachella Valley fringe-toed lizard than on any other desert species unique to the Coachella Valley. He wouldn't lose time conducting further research to bolster his report to the California Department of Fish and Game. Dr. Mayhew decided to get the Coachella Valley's own sand-diving lizard declared an endangered species.

Aztec kings wore ornaments of eagle claws and tiger bones to render them swift and invincible in battle. In a sense, Dr. Mayhew too chose an animal charm, an aid in his battle to save desert. Now, afterwards, people find it ironic that a four-inch-long creature could impede huge developments and infuriate influential businessmen. Actually, the lizard was a strength-giving ornament with which Dr. Mayhew, joined by other environmentalists, fought for territory—an ecosystem in the desert wilderness.

IV.

LISTING THE LIZARD

Spurred on by his friend's unofficial advice, in 1972 Dr. Mayhew petitioned the California Department of Fish and Game to classify the Coachella Valley fringe-toed lizard as endangered.

The state of California and the federal government each have an endangered species list, and each uses slightly different terminology to classify species on each list. California differentiates between "rare" or "endangered" species, while the United States Fish and Wildlife Service differentiates between "threatened" or "endangered" species. Bureaucrats could argue about fine distinctions which the terms reveal. Essentially, a "rare" or "threatened" classification means the government thinks a species faces a lesser peril of extinction than if either government had declared it "endangered." Getting placed on an endangered species list is a good thing for a species, for it will then receive government money and protection. For this reason, conservationists want to get a species listed, and don't worry overly about its classification (except that "endangered" sounds better and conservationists generally would like to shoot for that title).

Dr. Mayhew believed his research proved that the lizard faced extinction unless the state moved quickly to save the lizard's habitat. He believed the bureaucratic wheel leading from his request to a preserve would turn smoothly. The CDFG would add the fringe-toed lizard to the endangered species list, and thereby free public funds with which to buy permanent habitat for the imperiled reptile. The lizard would share its good luck—a preserve— with the rest of desert wildlife. Most people had never heard of a fringe-toed lizard and so Dr. Mayhew did not conceive that anti-conservationists, all boiled up over a lizard, would accuse conservationists of employing the power of government to ruin citizens' chances of prosperity.

"Well, the state turned me down," said Dr. Mayhew. "There was a funny psychology at that time that having to list the lizard to create a desert preserve somehow implied the department hadn't been doing its job of protecting wildlife.

"I said, 'Since you claim it isn't endangered, you'd better do something to prove to me that that's so.' Finally, the department said, 'You can't protect the lizard anyway without habitat.' I answered, 'Why don't you buy habitat?' Well, they couldn't get at the money unless the lizard was listed....

"It was a *Catch-22* situation. But the one thing I'm good at is rattling cages. I kept haranguing them. When my friends told me $200,000 was the most the department had spent at one time for habitat, I went after them for $200,000. At least they paid for a survey to study the situation of the fringe-toed lizard in the Coachella Valley— probably just to get me off their backs."

Sidney England and Steven Nelson finished the survey, "Status of the Coachella Valley Fringe-toed Lizard," a year after Dr. Mayhew approached the CDFG. Sid England said they tried to render an accurate overview of the Coachella Valley—from wild dunes to resort development. First, they described the land, the valley's "mosaic pattern"[21] of flat sand fields, darker, undulating sand hillocks piled by wind behind creosote bushes, and bigger, white dunes crowned with the half-swallowed branches of grey-green mesquite. Then England and Nelson explained where in this sand mosaic the fringe-toed lizard dwells. They examined the sand particles of the lizard's habitat and sent them to soil laboratories to clarify which sands the lizard needs.[22]

Leaving the wild regions, they turned to city general plans to find which areas of the lizard's sands were destined for homes, golf courses and cemetaries, horse trails and shopping centers.[23]

They read the development controls of the county and found the county had designated some areas as "open space," meaning no shopping centers, but light agriculture, or meat packing plants, or public utilities. England and Nelson cautioned that the valley cities could someday annex this open space and rezone it for heavier development.[24]

They also warned that windbreaks could change blowsand habitat into hardened areas by preventing the deposition of new sand unless measures were taken to protect the desert.[25] People don't like blowing sand. It stings eyes, dirties swimming pools, and makes roads unsafe. People strive to bring under submission the wind that hurls into the valley the loose sand which creates a living desert habitat.

England and Nelson mentioned that which cannot be analyzed in a laboratory or looked up in a city code—the concern of many residents about the valley's worldwide popularity changing a peaceful resort into a crowded, noisy, smoggy metropolis from mountain to mountain. Without being concerned about the development which could obliterate a kind of sand lizard, people had begun to be alarmed by development obliterating a kind of lifestyle.[26]

(Preserving this slower pace of living becomes intertwined with the fight for conservation so that conservationists, if politically savvy, may nourish an alliance with people who don't care about lizards, but just don't want any more Los Angelinos to move in.)

In the conclusion to their survey, before they could make a recommendation to the state, England and Nelson had to consider word by word the distinct state and federal differentiations between stages on the way towards a species' extinction.

California law says a "rare" species is found in such dwindled numbers that *changing its habitat* may jeopardize it. An "endangered" species is in *immediate peril* of extinction.[27]

The federal Endangered Species Act of 1973 says a "threatened" species may be in jeopardy *in the future* as we foresee it. An "endangered" species is, on all or only parts of its habitat, *now* in peril.[28]

England and Nelson felt they could not say to the CDFG that the lizard was found in such dwindled numbers that disturbing its habitat could decrease its chances for survival. Any warm day, fringe-toed lizards swarmed over the sandy areas still undisturbed, and actually, a lot of these lizard sands remained. The biologists conjectured that until the habitat had shrunk by twenty percent, the fringe-toed lizard couldn't be called rare. By this reasoning, England and Nelson certainly could not say the fringe-toed faced immediate peril and so classify it as endangered.[29]

Instead, they included in this report to the state a recommendation that the *federal* government list the species as threatened. They said the development foreseen in Palm Springs's future would jeopardize the survival of the fringe-toed lizard.[30]

Also, they urged the CDFG to not wait for the lizard to fit into a category. By that time, its immediate peril of extinction could mean the more profound death of a whole desert ecosystem. They urged the department to establish a desert preserve and they even suggested a site—south of Ramon Road, against the Indio Hills, north of Interstate 10, west of Washington Street (See Map I)—Dr. Mayhew's spot exactly.[31] England and Nelson's recommendation for a site delivered a punch: it implied, on state record, that to protect an unusual reptile fated for having adapted to a warm clime desired by humans required governmental creation of a sand preserve. No other path existed.

The CDFG dilly-dallied about publishing the study. Dr. Mayhew inundated the department with letters. He wrote to anyone else who might harbor interest in what he had had to say to the CDFG. In May 1977, he wrote to the Desert Protective Council, an organization founded to safeguard deserts of the southwestern states, for help in "shaking loose" the study: "The England-Nelson survey was finished nearly two years ago. Their report was submitted to the California Department of Fish and Game over a year ago, and the final draft was completed over nine months ago....I am getting the feeling that someone in Sacramento is 'sitting on the report' for some reason."

Drafts of the report ping-ponged between England and Nelson, and top people in the CDFG. Finally, the agency edited and printed the report. Sid England said his and Nelson's initial draft had used much stronger language than that which the department printed, and to which it attached an extra editorial England and Nelson didn't agree with. Furious, the authors demanded that distribution of the report stop. The department compromised and let England and Nelson write a list of corrections to the CDFG version and to slap it on the front of the report.

The department had printed that the lizard not be protected until *one half* of its present domain had vanished. England and Nelson corrected this to read "We recommend that *U. inornata* be considered threatened under the meaning of the Federal Act and rare under the State Act by the time the existing suitable habitat of 311 km squared (120 miles squared) is reduced by an additional 20%."[32]

By the time the department reprinted the report, Dr. Mayhew had already despaired of the state and had sent to the USFWS the original England/Nelson survey which both recommended a federal listing of threatened, and the creation of a desert preserve. In September 1977, the *Federal Register* advertised with noncommital, bureaucratic style, that the Office of Endangered Species was thinking about studying whether or not to propose to list the lizard and to define its "critical habitat," the area considered essential for the survival of a protected species (yet not its entire geographic range).

Coachella Valley residents reacted lividly. "What the hell has the fringe-toed lizard done for me and humanity lately?" said one man to a newspaper reporter.[33] Frankly, developers, residents, businessmen, realtors, and city officials were frightened of what the lizard could do, or more precisely, what the federal government would do if it decided to aid a lizard camouflaged to not be seen in the sand—except by a flash of movement. Dr. Mayhew had flashed the lizard to the government and the movement exploded—in the Coachella Valley a long battle of neighbor disgusted with neighbor and the fringe-toed lizard.

From neighbor to neighbor, businessman to realtor, homeowner to shop owner, from newspaper to newspaper, the fear traveled: people believed the federal Endangered Species Act prohibited the use of federal money in any way which jeopardized a federally protected species. If Joe Smith wanted a federal loan in order to build right smack in the critical habitat of the lizard he might find he would have to dig up money elsewhere.[34]

A specter of the Palm Springs area cut off from federal loans and sunk in the depths of a recession gripped residents. One newspaper reported the USFWS was "...proposing that virtually the entire Coachella Valley be made into a vast wild life preserve...." The local water district opposed a listing, and announced that an "endangered" lizard would wipe out federal funding for flood control. (Some floods rush from the fringe-toed lizard's habitat towards human homes.[35] A flash flood that had swept through the town of Thousand Palms, cutting under streets, and dumping mud in living rooms, rankled fresh in people's memories. They shouted, "We need flood control"—and no lizards.)

Actually, the "entire valley" would never be deemed critical habitat. The lizard does not live throughout the whole valley. But by that point in the battle many people didn't care if there were no lizards anywhere near Palm Springs. They said that even if the fringe-toed lizard were exterminated here, plenty others abound throughout the southwestern U.S. and northern Mexico. Dr. Mayhew countered that the Coachella Valley fringe-toed lizard has evolved as a distinct species from other fringe-toeds. Then, biologists argued whether the Coachella Valley fringe-toed lizard should be classified as a separate species, or as a subspecies.[36]

"Species" defines a group of individuals *every one* of which possesses definite, unvariable characteristics. Yet, because natural barriers such as rivers or mountains cross the geographic range of a species, small groups within the species become isolated. These smaller groups within a species evolve shared physical variations, and the *average number* of members of each small group *generally* displays the variations (while also displaying the definite characteristics of the species). Biologists label these smaller groups as subspecies.[37]

One way to determine whether a group forms a species or only a subspecies is that individuals from separate species do not interbreed, while individuals from various subspecies will.[38]

The Coachella Valley fringe-toed lizard has interbred in captivity with fringe-toed lizards from other areas.[39] To some zoologists, this classifies the Coachella Valley fringe-toed as a subspecies. Others disregard interbreeding in captivity and claim it doesn't prove what occurs naturally. These zoologists maintain that the Coachella Valley fringe-toed lizard forms a distinct species.

Some local biologists feared that to try to protect a mere subspecies would rouse the public's wrath unnecessarily and then sour people against conservation. Dr. Mayhew insisted that, species or subspecies, no lizard like the Coachella Valley fringe-toed occurs anywhere else in the world, and it would be a poor thing if people chose to destroy it. Dr. Al Muth, a biologist whom Dr. Mayhew hired to join the fight for the fringe-toed lizard, contends the argument never jeopardized the lizard's prospect of being added to the list since the Endangered Species Act protects both species and subspecies. But the debate split conservationists who argued the political side effects of saving a subspecies. In the technical dispute, biologists forgot that they could be saving an environment, not only a subspecies of lizard.

Other disagreements emerged among the environmentalists. Some resident conservationists bristled at the suggestion the valley needed another preserve because many years ago, desert lovers founded the Living Desert Reserve. The reserve has grown into a nature museum with buildings for school programs, and a bookstore and giftshop. Various gardens represent different deserts of the United States. Nature trails web the 1,200-acre reserve. Longtime supporters of the Living Desert Reserve dreaded competition from a new preserve for money and local interest. Perhaps the proposal for another preserve offended them and implied the Living Desert Reserve hadn't done enough for desert wildlife. They pointed out that both the Living Desert Reserve and also UCR's Deep Canyon Research Center already protect raw desert. Dr. Mayhew said these reserves do not protect the loose blowsand habitat of the Coachella Valley fringe-toed lizard.

Displeasure with competition arose among supporters of the Deep Canyon Research Center. Jan Zabrinsky, the biologist UCR hired to take care of the Deep Canyon Research Center, had engrossed himself in the lizard fight as director of the Coachella Valley Ecological Reserve Foundation (C-VERF), which Dr. Mayhew had organized to work for a blowsand preserve. One influential supporter of both the Living Desert Reserve and the Deep Canyon Research Center disapproved of Jan Zabrinsky's involvement in C-VERF and said that Zabrinsky, and later Al Muth, who took over Zabrinsky's Deep Canyon job, "should have had enough to do keeping the Deep Canyon Center in order and should have concentrated on that," rather than submerging themselves in the local lizard affair.

Dr. Mayhew's preserve couldn't possibly compete with either the Living Desert Reserve's nature displays and long membership list, or the Research Center's scientific facilities. Those environmentalists who believe educational goals legitimize conservation don't understand a lot of people whom the pedantic approach to wilderness leaves feeling suffocated. The new preserve would appeal to these people because it would neither explain the desert to the public, nor serve researchers solely. Instead, for poets and photographers, for walkers and thinkers, for all who esteem an undisturbed and untranslated wilderness, the new preserve would safeguard the vivid Colorado Desert as God made it.

In spite of disagreements over what constitutes a justified use of wilderness, most conservationists want to hold onto all land possible. They like to "keep things just they way they are"—no urban expansion—and they find it difficult to comprehend that to the realtor promoting a resort, or the little guy down the block who just wants to get ahead and be financially comfortable for once, this freeze constitutes the antithesis of progress and prosperity. To them, it's unbusiness-like to purposefully set aside, in a sort of time capsule, an area where sand blows like whirling dervishes. Sand's a nuisance in pools, on roads....To block off an area labelled critical habitat means an already small airport can't be extended into this critical habitat. To put a lizard above humans and wave goodbye to federal funding cramps the breathing space a new sewer plant needs.

The Indio Chamber of Commerce (followed by its Palm Desert counterpart)[40] issued a proclamation calling the cities together to protect the local people's rule by preventing the federal government from listing the lizard: "Whereas, designation of the area as a critical habitat for the Coachella Valley fringe-toed lizard without reevaluation of those portions of the designated area which could not sustain the lizard, and without evaluation of the economic impact on the above mentioned people ["every property owner, realtor, developer, builder, financier, grower, packer, shipper, and business owner in the Coachella Valley."] would spell economic disaster for the Coachella Valley...."[41]

Yet everyone harbors his own idea of home and city, and how they should be cared for. A school child wrote to the USFWS that "the fringe-toed lizard may be just a lizard, but how would you like it if you were sitting at home and some big old tractor came and dug you and your home up?"[42]

A Rancho Mirage city councilman agreed with the little girl. He advocated for cities to plan ahead for the day the entire desert floor would be developed. A preserve would save "a segment of the original desert floor" for "our children's children."[43]

Dr. Mayhew heard through his friends that the CDFG had recommended to the USFWS that the federal agency *not* list the lizard. Dr. Mayhew wrote to the state secretary for resources, "I am rapidly coming to the sorrowful conclusion that unless a vertebrate species is hunted or fished, it has little chance of being adequately protected by the current Department of Fish and Game."

In spite of the state's suggestion, the USFWS gave the lizard a chance. The *Federal Register* of September 28, 1978 published the actual proposal to put the fringe-toed lizard on the federal endangered species list, and to define parts of the Coachella Valley as the lizard's critical habitat. "We were so lucky," said Dr. Mayhew, "that the proposal came out then because two days later Jimmy Carter signed a bill amending the Endangered Species Act. If the lizard hadn't gotten in the *Register* before that, it never would have had a chance to even get proposed later in the *Register*, not with all the subsequent feeling against it."

This amendment to the Endangered Species Act required that the USFWS, before deeming an area critical habitat, prepare a report determining the *economic* impact which the environmental restrictions of "critical habitat" would inflict upon that area. The proposal to designate a part of the Coachella Valley as the critical habitat of the fringe-toed lizard had to be withdrawn while the USFWS staff studied the situation and wrote an economic impact report. Local headlines read "Thousands of Rare Species Lose Federal Protection,"[44] and falsely reported that the *lizard* had been dropped from the endangered species proposal list. "Once on that proposal list," said Dr. Mayhew, "the lizard never was taken off."

In 1978, when President Jimmy Carter signed the Endangered Species Act amendment, "developers laughed," said Dr. Mayhew. They said "no problem" to proving that a critical habitat designation would crush the valley economy. They gloated that the lizard hadn't a chance because the riches to be gathered from the golfing, shopping, eating, drinking, dancing, and marketing of tourists gleamed clearly an economic right of humanity. People are entitled to use the resources of the world right now, to borrow money for a housing project and to build it where they want, to sell it, to collect a profit. One resident, Virgil Kraft, wrote to a local newspaper that "lizards and odd fish" may interest scientists, "but cannot fuel a car, build a home, add one bit of beauty to a drab neighborhood, or keep sensible people from getting sick to the stomach."45

Such sensible people pooh-pooh conservationists as emotional and impractical. True nature lovers admit that their souls fuel their drive to preserve. The messages of coyotes ringing from hill after hill around the valley resound as reason enough to save wilderness. The defenders of nature think how awful it would be to live in a world where everything—buildings, mazes of streets, art—reflects only the ideas of another human mind. In such a world exists no bedrock from which the individual soul feels it was birthed, or upon which it consoles itself for being human. And being human, individuals crave various lifestyles. Watching lizards dive in sand thrills an admirer of nature like a game of golf exhilarates the weekend linkster. He demands his right to be happy playing his game in the sun on his own kind of course. He wants to use the earth's resources *right now* by savoring the wilderness on a backpacking trip. He sounds the most ridiculous to the anti-conservationists when he counters that as far as rights go, all wildlife, lizard to lion, has a right to live where it has perfected the art of living.

Conservationists battle sounding ridiculous and impractical. They seek economically steady arguments and publicize that if we thrash about the earth, breaking links in the life chain, our own existence will hang as precariously as that of this lizard. We receive air from the earth, and food from the earth, our shelter, our medicine, our clothes, our toys. We had better protect our earth as she was given to us so that later we won't find we've thrown away something we want, or we desperately need. What living link can our elected governors rule be dropped, judged insignificant, dispensible?

The arguments drew fighters to opposing camps. The Riverside County Board of Supervisors voted to discourage the USFWS from designating the Coachella Valley as critical habitat.46 Realtors met to discuss how to squelch the listing.47

Conversely, C-VAG (the Coachella Valley Association of Governments), an organization of representatives from some valley city councils and from the county, and which strives for long-term, cooperative planning, voted to support a preserve in March 1979. The representative from Rancho Mirage, a biologist, suggested a compromise—if together the cities established a big dunes reserve, the lizard would automatically be protected and so the quest to list the lizard could be abandoned. Free of hassle, free of environmentalists' surprises, developers could build up the valley around the preserve.[48]

A group of developers and realtors called "Friends of the Fringe-toed Lizard" (The conservationists' camp hotly shouted that with friends such as these....) liked the idea of conservationists' agreeing not to seek governmental protection of the lizard. The Friends said people should solve the problem without governmental interference. The Friends proposed to locally organize the establishment of a preserve. Their price would demand the lizard not be listed and that no restraints inhibit valley development.[49]

"How could we trust them to establish a preserve?" Dr. Mayhew asked. "Once the government was out of it and the lizard removed from the race to hit the list, they'd have no reason to do us a favor. And the state would say 'Forget it' if we proposed the lizard again. They'd tell us to make up our minds. So we kept pushing."

Sid England shot aerial photographs which showed that less than 100 square miles of habitat remained as compared to when the CDFG published his "Status of the Coachella Valley Fringe-toed Lizard." The criteria he had set up in the CDFG study had been met for the lizard to be listed by the state.

The CDFG held a public meeting in Palm Springs on June 16, 1980 about the lizard listing.[50] Angered residents thought the state representatives came to the meeting with their minds made up. George Whitney, the real estate broker for Justin Dart (who owned the land where Dr. Mayhew wanted his preserve), said, "The CDFG advertised the meeting in the second to the last page of the *Riverside Enterprise*, not the local paper, and held it in the middle of July." It's an old Palm Springs political trick to hold a spark-and-tinder meeting or an election in the middle of the summer. Anyone who has a choice has usually vacated the desert by then.

Eleven days later, at the California Fish and Game commissioners meeting to decide the fate of the fringe-toed lizard, the lizard issue was first on the agenda, at 8:30 in the morning, "...which is okay," said George Whitney dubiously, "but you know where it was? Lone Pine, California, way up by Bishop, population 2,000."

Justin Dart chartered a plane and sent his lawyers to Lone Pine to battle. "At great expense," added Whitney, "but the state passed the listing anyway."

Dr. Mayhew said the California Fish and Game Commission, which votes on listing proposals, definitely did not decide until the last minute, but that the Friends of the Fringe-toed Lizard cut its own throat. The Friends had threatened Dr. Mayhew and his supporters, saying that developers owning land in the proposed preserve would immediately plow the desert for farm fields if the conservationists didn't agree to the Friends' local solution, and abandon the fight to list the lizard.

Either the Friends didn't give themselves time to carry out their threat or didn't anticipate the gall of their opponents. At the meeting, five commissioners wielded the fate of the lizard. Dr. Mayhew knew two would vote for the listing and that two would vote against; the chairwoman's vote hung undecided. The meeting rules permitted each side to speak once. Osborn Brazleton of the Friends of the Fringe-toed Lizard spoke first. Then Steve Nicolai from Dr. Mayhew's group told the commissioners of the threat. Dr. Mayhew said, "The chairwoman responded, 'That sounds like blackmail to me.'"

After years of dragging its feet, the CDFG decided that the phenomenal expansion of the Palm Springs resort area jeopardized the Coachella Valley fringe-toed lizard, and on June 27, 1980, declared the lizard endangered, bestowing the highest degree of protection under California state law.[51]

Dr. Mayhew was jubilant. A state listing strengthened his request to the USFWS to add the lizard to the federal endangered species list. A federal listing would open bigger stores of money, and with greater power protect the lizard from habitat obliteration by development

"I copied the minutes from that triumphant CDFG meeting, which told all about the blackmail attempt, and sent them right off to Washington," said Dr. Mayhew.

Having finished the economic impact report and having reproposed the Coachella Valley as critical habitat, the USFWS faced whether or not to support the CDFG and protect wildlife, or to let development boom without restriction. A spokesman for the USFWS emphasized that the purpose of the listing would be to free funds for habitat acquisition, not to block government housing loans, but when newspaper reporters insisted he explain more, he admitted no one knew if the federal courts would deny loans for low income housing on critical habitat.[52] The law hadn't yet been tested.

Palm Springs has not been developed with the dwelling quarters of lower income families in mind. At six o'clock in the morning, hotel maids, waiters, janitors, gardeners, and construction workers pack the roads leading to the valley, driving close to an hour to get to work. As the resort expands and tourists require more services, the workers will need places close by to live, such as the Thousand Palms area, where Dr. Mayhew wanted a preserve. Opponents to a federal listing fumed that if the desert around Thousand Palms were deemed critical habitat, the cost and time to prepare both environmental and economic impact reports would eliminate the feasibility of low income housing projects.

Dr. Mayhew kept fighting. He said he found out about a meeting in Washington between Coachella Valley developers and representatives of the Bureau of Land Management, the USFWS, and Jerry Lewis, the congressman representing the Thousand Palms area. The meeting took place one day after the closing of the public comment period. The meeting should have taken place when Dr. Mayhew could have attended, and defended the listing of the lizard and the declaration of a critical habitat. He accused those at the private meeting of having discussed a land exchange between the BLM and developer Justin Dart, whose condition for an exchange required that all the Dart holdings be exempted from any critical habitat designation. Part of Dart's property comprised a big chunk of Dr. Mayhew's preserve in Thousand Palms. Dr. Mayhew redoubled his letter war.

He wrote to Congressman Jerry Lewis and sent copies to the director of the USFWS, all five Riverside County supervisors, and two other congressmen and one senator for California. He urged Lewis to support the listing, saying that developers blocking the listing were merely "postponing the inevitable. As more lizard habitat disappears under the developer's equipment, the greater will be the outcry to preserve what is left." Dr. Mayhew insisted that the problem could not only be local ever. All Americans lose when wilderness disappears.

Only Congressman George E. Brown Jr. answered. Dr. Mayhew credited Brown with pushing through the federal listing. Because Congressman Lewis wanted a dam built to control floodwater in the Whitewater Canyon above Thousand Palms, he feared a federal listing would wipe out funds for the dam.[53] Dr. Mayhew said that Brown, an older congressman and a Democrat, "defanged Lewis," a first-term Republican in a Democratic administration.

On September 19, 1980, Dr. Mayhew answered his telephone. A newspaper reporter asked him, "How do you feel?"

"What are you talking about?" asked Dr. Mayhew.

"Didn't you know? The federal government declared the fringe-toed lizard threatened today, and part of the Coachella Valley its critical habitat."[54]

A threatened, as opposed to endangered, listing didn't rank second best because, by naming the lizard a threatened species, the USFWS had decided the lizard was "likely to become an endangered species within the foreseeable future...."[55] The USFWS, therefore, had to develop a "recovery plan"[56] to fetch the lizard back from that precarious state of existence, perhaps through buying habitat. Also, by naming parts of the Coachella Valley critical habitat, the USFWS could stop development on expensive land in the Coachella Valley.

The city of Rancho Mirage didn't like having critical habitat, land with no future in development, within its boundaries.[57] Dr. Mayhew says, "Imagine, they threatened to sue the Feds!" But the city had little to fear. Gail Kobetich, director of the Sacramento Office of Endangered Species of the USFWS, admitted later the USFWS stalled about enforcing the ESA (Endangered Species Act). The ESA forbids anyone to "take" a federally protected species.[58] The law clarifies "take" to mean to "harass, harm, pursue, hunt, shoot, wound, kill, trap, capture, or collect,"[59] in other words, to keep your hands off the Coachella Valley fringe-toed lizard. That means the agency was supposed to tell people with an endangered species residing on their property that they couldn't do anything with the land except picnic there and admire the protected species. Without being able to suggest a solution other than stopping development entirely, the USFWS didn't want to prosecute developers destroying lizard habitat.

At that moment, developers weren't frequently violating the law because they weren't developing. An economic recession and a slowdown in second home buying had stopped development. Even so, when the CDFG tried to buy lizard habitat with funds allotted endangered species, no one would sell. Everyone swore the Palm Springs building boom would explode again, and that property owners would get rich if they waited to sell to a developer. Dr. Mayhew's plans for a preserve sank in a quagmire of unstable laws and economics.

The USFWS's recovery plan (as required in the ESA) for the lizard called for two or more preserves without specifying how big or where they should be. The USFWS held a public meeting in Palm Springs to locate sites. In the 1970s, Dr. Mayhew had outlined an oblong shape running the same direction of the valley—east and west—so that wind blowing in through the San Gorgonio Pass would supply sand from east to west. The USFWS had to satisfy itself that preserving this area meant a big chance for the lizard's survival. Also, the USFWS had to decide which areas it feasibly could protect or acquire.

The USFWS excluded the lizard habitat north and west of Ramon Road because many people owned a myriad of small parcels there which would confound the purchase of a big hunk of habitat. (See Map I)

At the Palm Springs meeting, Don Weaver, the design engineer of the San Bernardino office of the California Department of Transportation who had conducted the department's 1972 study, testified that, because of existing structures and future planned development, in five to twenty-five years the land south of Interstate 10 would be so shielded from blowsand that it would be uninhabitable for the fringe-toed lizard.

The USFWS wanted to ensure that soft sand would continually blow over the lizard preserve. But if it relied on wind, the preserve could be doomed. Who could say the whole desert, except for the preserve, wouldn't one day be developed? From where then, would sand come to keep the preserve from hardening and dying? So that the preserve would include Thousand Palms Canyon, from which floods would supply sand, the USFWS swung the boundaries of Dr. Mayhew's original preserve from the east and west around to the north and south. Even if the valley became a metropolis, development couldn't stop rain and floods.

Desert realtor George Whitney scoffed at the sand source dilemma and said the USFWS changed boundaries "because the boundaries just happened to fit neatly into Justin Dart's property since it's a lot easier to buy from one owner than a bunch of little guys."

Whitney felt the government essentially forced Justin Dart to sell by deciding it wanted a preserve around Thousand Palms Canyon. Dart owned a huge parcel—not only the sand fields Dr. Mayhew wanted, but also Thousand Palms Canyon and its desert oases filled with native California fan palms.

In 1905, a cattle rancher from Hemet, looking for summer water for his herds, listened to rumors of an oasis

in the desert. Arriving in the Coachella Valley, he chanced
into an old friend, Alkali Al, living among the palms and
pools of Thousand Palms Oasis. In exchange for eighty
acres, including the oasis, rancher Louis Wilhelm traded
Alkali Al a pair of mules and a wagon. Alkali Al rode off,
leaving the solitude to settle over Louis Wilhelm.[60]

His son Paul, captured by the seduction of water in
the desert, of great palms whose dried skirts and green hats
hiss in the wind, said when he was fourteen, "Dad, someday, just leave me here alone." Beneath the palms he listened to the chanting of Indians healing their sick in the
water. But his father said, "You have to get an education."

After spending World War II "following Patton
across Europe," Paul Wilhelm came home to Thousand
Palms Oasis, and the spread of the city towards the Indio
Hills and the oases shocked him. Glamor people from
Hollywood had pounced upon the desert. They stayed a
week or so. Then, tanned and relaxed from the swimming,
the horseback riding, and the open air, left. Today, old-
timers remember those as having been blissful days in a
small town. Even then, Wilhelm foresaw the Palm Springs
that grew to sprawl all the length of the valley, split into a
myriad little cities and fattening towards the Indio Hills.
He called together the owners of oasis and canyon lands
and forecasted the gigantic development to come. They
signed a consultation agreement so that one owner wouldn't
sell without talking it over with everyone else. None of
them could afford to donate the land to the state park service, but Paul Wilhelm believed they could encourage nice
development if they all offered the canyons and groves and
bluffs as one big piece.

In 1972, when Justin Dart bought much of their
land, the former owners and he shook hands that he would
not parcel the property out in small plots. Dart had bought
much more than the oasis—19,000 acres comprised his new
property, encompassing the hills above the oasis and the
sandy desert below. (See Map III) For ten years he held on
to it without developing. Perhaps he was waiting for the
fringe-toed to lose its punch. Some people say Dart's corporation was doing better selling plastic kitchenware than

real estate and that the SEC wanted the Dart Corporation to divest itself of realty holdings.

In any case, Dr. Mayhew's preserve and Paul Wilhelm's oases remained undeveloped when Dr. Charles Allard of Canada paid $13 million in cash for Dart's 19,000 acres in 1972. Dr. Allard knew about the fringe-toed lizard, but, according to his realtor, George Whitney, "never thought they'd politically railroad it like they did."

Dr. Allard's plans split the area into three sections.

The sandy land below the oasis, with water and power lines and roads already established, would nurture lush fields.

For the hills, Dr. Allard dreamed of a complete, self-sufficient solar city rising above the desert floor. Regular gasoline-dependent cars would be parked outside the city gates. Inside, residents would travel in solar-powered vehicles.

"The area is perfect for a solar resort," said realtor Whitney. "It gets four hours more sun than the rest of the valley. The view is spectacular of Mount San Jacinto, Mount San Gorgonio, and it overlooks all the cove communities."

In the oasis and canyon lands, Dr. Allard wanted a park. "All natural." asserted Whitney. "Maybe they'd enlarge the oasis dam so the lake would be bigger. A few more trees planted, and the RV trailers hidden among them."

Dr. Mayhew rattled the Bureau of Land Management's cage to do something fast about saving the preserve. The BLM suggested a land trade.61 George Whitney told the story: "I had a client who was going to buy from Allard and trade with the BLM. So he hired an expensive appraiser to cover both his land and the BLM's, which was spread all over Southern California—five acres here, five there. Way out in the mountains too, impossible places. He spent months, and my client spent lots of money. Then the BLM appraiser spent one or two days down here and said our land was worth way less than theirs."

Dr. Mayhew agreed. "The figures were ridiculous."

No one was selling, and bulldozers kept biting chunks out of the lizard's habitat.

Jan Zabrinsky, the director of C-VERF, was leaving the Coachella Valley to study environmental law, and Dr. Mayhew had to look for the new Deep Canyon Research Center manager. As he interviewed applicants he watched out for someone who'd also take over C-VERF.

Dr. Al Muth was living in Colorado when he got a phone call from a friend that UCR needed someone in Deep Canyon. "Jobs for doctorates are difficult to come by," Muth said, "and this was an alternative to the academic life anyway."

He first met Dr. Mayhew in Riverside. The pysician detected latent fire and fury behind this biologist's gold-rimmed glasses and beard, and inquired if Muth would take an interest in local environmental issues. So Muth assumed his job would include lecturing at museums, and explaining environmental impact reports at city council meetings—nothing strenuous.

"I knew what the fringe-toed lizard was because I'd been in the Coachella Valley before, studying the desert iguana," he said. "But until the interview, I hadn't heard *anything* about the problems it was causing."

In January of 1982 Muth and his wife moved to a little house in Deep Canyon that used to be a stables' tack room. Other biologists studying in the canyon for a few months or so live in two house trailers. The cement block labs smell of preserving fluids. "We may not have much," said Al Muth, "but we've got the best view in the valley." The eye sweeps down the brown canyon that fans out to the dark pattern of resorts and fields upon the desert floor. No noise of traffic floats up to disturb the stillness in January when the air crackles with spots of crystal and the researchers dwell as kings above the city. Dr. Mayhew gave Muth two weeks to enjoy the peace and then called him.

"The next meeting regarding the fringe-toed lizard is next week," he told Muth. "I think you should be there. By the way, we've decided you are director of the Coachella Valley Ecological Reserve Foundation. It's no big deal. You may be making some phone calls." Muth laughed, "I guess I have made a few...and taken a few."

Quickly, he realized his job meant for him to be rabble-rouser. He said the state endangered listing had clout, but "the state is a slow-moving, conservative bureaucracy. Every little thing has to go through committee after committee. To make a bureaucracy listen, you need a local person to get up and scream, and make it enforce its own laws."

One day that March, he drove down the highway in Palm Desert, directly below Deep Canyon. A stand of big, twisted creosote bushes there in the desert had intrigued him for weeks. He had kept meaning to go take a look, but what with moving in and figuring out his job, he just hadn't. That day, Muth stopped the car in horror. Desert had been transformed into chewed-up dirt—the universal blank of all construction sites—ready for the new Palm Desert Mall.

"I was shocked. That's when I really understood that if something wasn't done fast to preserve the desert it would be too late."

Gail Kobetich, the USFWS' director of its Sacramento Office of Endangered Species, said that without Muth, the effort to save the fringe-toed lizard would have been stymied. Developers could say "We don't care what the ESA says," and continue developing. If no local person cared either, the USFWS would have a hard time insisting the law be obeyed. But with a local person screaming to the USFWS, the agency got the lever it needed. Kobetich said, "Cities saw there was somebody who would actually bring down the ESA. Al Muth was the main hammer. He virtually bludgeoned people into submission."

"Serendipity," said Muth often, crediting good fortune in the desert preserve battle. Luck, from the point of view of the conservationists, had brought Al Muth to the valley.

Muth himself said that because he was new to the valley, he was not discouraged, but rather eager to take on developers. "Fortunately, I was naive about how powerful developers are," he said. "Otherwise I might have lost my nerve."

He threw himself into the "battle of the biological surveys," which is how Paul Selzer, an attorney for the Sunrise Company, described this era in the Coachella Valley. Anytime developers proposed, either to the county or a city, a development in the lizard's critical habitat, they had to present a survey called an environmental impact report analyzing the effects which that particular development were expected to have upon the fringe-toed lizard's living environment.

"Developers would hire biologists who knew which side their bread was buttered on," Selzer said candidly, "and so write up reports saying there were no lizards on the developer's land.

"But there was stretching of the truth on both sides. Al Muth kept an eye on all the projects and would come back saying the reports weren't correct, but there's no way he could have gone out and checked all the information. In either case, the county board and the cities were making decisions using inaccurate information."

By refuting the surveys, Muth could delay projects, but he couldn't stop them. Instead, the economic slowdown reined in the local development. Sunrise Company, along with most developers, bided its time until a perked-up economy would flood the valley with buyers of second homes.

In late 1982, Muth's serendipity smiled again: Congress once more amended the Endangered Species Act. A friend studying environmental law sent him the amendments, certain portions paperclipped and outlined in red. Muth couldn't believe what he was reading. He called Gail Kobetich to make sure he understood: The revised ESA stipulated that if a company wanted to develop an area that was part of a species' critical habitat, the developer had to guarantee that the species would have evough of its critical habitat left to survive.

The writers of the ESA had assumed a developer would not storm his bulldozers onto the desert purposefully looking for lizards to squash. He merely annihilates lizards

58

by obliterating their habitat. (The ESA describes this action as "incidental take"[62] of a protected species.) It makes no difference if funding for activities destructive to the habitat derives from private sources. Harming a protected species is illegal.

However, the 1982 amendments set up conditions by which the USFWS could grant a property owner a permit to legally "take" a protected species. The property owner first prepares a "conservation plan" to ensure survival of, and even to replenish the species, such as leaving a portion of his land undeveloped, or paying money to the USFWS to buy habitat elsewhere. Thus he "mitigates" his destruction of habitat. He presents the plan to the USFWS, applying for an incidental take permit. If the USFWS believes the perpetrator of the conservation plan exudes dedication and that the plan will make up for loss of habitat, it grants the applicant an Endangered Species Act section 10(a) [ESA 10(a)] permit.[63]

"The amendment meant a way out for the Fish and Wildlife Service," said the USFWS' Kobetich. Previously, the agency could have told a developer, "Stop. You are harassing an endangered species." Yet what else was the developer to do? Not use his land and quietly accept yearly property taxes as a donation to environmental welfare? The new clauses offered an alternative to rigorous enforcement of the ESA.

After Congress amended the ESA, Kobetich, who believed development would soon eliminate the fringe-toed lizard in spite of its threatened status, took advantage of the way out. He went to the valley in March 1983 to work out a conservation plan. He talked to city governments, to the county planning department, and to the BLM. Propelled by the authority of federal law, he could tell these groups they had to work out a conservation plan. Said Kobetich, "The local governments were willing because they didn't want development stopped." In three to four months he had roughed out the conservation plan.

About this time, developers began dusting off blueprints in anticipation of the forthcoming surge of weekend home buyers. Sunrise Company started pushing its latest country club, Palm Valley, through the county planning commission. As usual, Muth examined the environmental impact report which the company had prepared for the county. The report appalled him.

"It misidentified major plants in the area of the proposed club," said Muth. "Worse, it copied verbatim UCR's lists of vertebrate animals which appear in Deep Canyon. The report claimed these animals were in the valley floor—they had *mountain lions* out there."

Even as he retold the story, he sounded amazed all over again. The biology in the report was so bad that Muth, ESA amendments in hand, went for the Palm Valley country club project like a bull terrier on the scent of something live and gamey. He knew the developers *had* to mitigate now if they wanted to build Palm Valley. He slapped the ESA amendment down on the tables of the county offices and mailed them straight to the president of the Sunrise Company, Bill Bone.

V.

THE LIZARD CLUB AND
THE BALANCE OF TERROR

Bill Bone called Sunrise attorney Paul Selzer. "What do you know about the Endangered Species Act?" he asked. Selzer read it and called Gail Kobetich of the US-FWS. "This is nuts," he said.

"The law's the law," Kobetich responded. "Call Al Muth."

So Selzer called Muth, who said that Sunrise could mitigate by buying 520 acres of prime lizard habitat—the same number of acres in the Palm Valley blueprint—or a prime, desert preserve.

"That was craziness," said Selzer later. "At $4,500 an acre, we're talking a lot of money. Besides, Al Muth couldn't tell me there were lizards on every foot of that 520 acres planned for Palm Valley, so why should we give him 520?"

Instead, Sunrise offered Muth's group, the Coachella Valley Ecological Reserve Foundation, $25,000.

"I saw the check," said Dr. Mayhew, "but Al wouldn't take it. Twenty-five thousand dollars wouldn't buy a big enough preserve. And once he allowed that kind of exception he knew he'd lose his strength through the ESA and never get any mitigation out of another developer again.

"I hear Selzer offered the CDFG $25,000 too—probably the same check."

Said Selzer: "I told Bill Bone, and I still believe this, that if we gave them $100,000 it would take care of the lizard problem. Muth knew he could no way get a $2 million preserve out of us. With the $2 million with which they wanted us to buy a preserve, we could have run them

61

out of business with a hell of a suit. I give Bill Bone credit for smoothing out this whole mess."

Bone knew that unless the lizard dilemma was untangled, he'd have a battle over every one of his future developments (especially with Al Muth around) because all of his company's major plans aimed for building smack in the realm of the fringe-toed lizard. Bone wanted to smooth the way for his company's developments, and, said Kobetich, to preserve Sunrise's reputation in town as nice guys. Instead of battling ecologists and their law, Bone resolved to make plans for the lizard too.

"What developers really hate is surprise," said Dr. Mayhew. "If they are prepared to pay for ecological mitigation, they merely pass the cost on to the buyers. But they want to know what's ahead and be able to plan."

"If a little developer with no long-range plans in the valley had been our opponent," Muth said, "we wouldn't have gotten anything and what's happened now wouldn't have happened. Serendipity—Sunrise asked for a building permit right after we found out about the amendments. If I'd known how big Sunrise is, though, I might have gotten scared off."

As president of the Sunrise Company, Bone possessed the equipment to upheave much more than sand dunes: influential people listen to a big developer. Bone decided he wanted to loosen development from the snarl of "take" and "mitigate" by protecting the lizard. Sunrise attorney Selzer said that, in total, counting outright donations, and also business expenses, "Bill Bone has spent lots, lots more than if he'd written a $100,000 check."

One month after Muth had mailed those photocopied pages of the ESA to Sunrise's headquarters, Bone invited all the protagonists and antagonists to lunch at his company's Monterey Country Club in Palm Desert: Dr. Mayhew, Al Muth, Paul Selzer, Gail Kobetich, and Congressman Al McCandless, among other representatives from the BLM, C-VERF, the CDFG, the USFWS, and the county Supervisors' office attended.

Said Kobetich, "That meeting and ones after were fairly contentious because of mutual distrust. Developers

62

distrusted the motives of the Fish and Wildlife Service, which in turn distrusted the developers' sincerity in solving the problem. After we got over that, things became relatively easy." Eventually, the group dubbed itself the Lizard Club.

Selzer described the meetings as "negotiations between various interests within various interests for a biologically defensible solution which was economically and mechanically feasible, and politically palatable. Ha!"

At the first meeting in the Monterey Country Club, Kobetich presented the conservation plan he'd been forming. He said that in order to mitigate, the developers would have to give money, or land, or both just for a lizard. In turn, environmentalists would have to be happy with *that* and not demand all the habitat in the valley. One way a developer can mitigate the effect of his development upon the environment is if he preserves a segment of his property in its natural condition. With many developers mitigating this way, the resulting habitat preserves would comprise small plots isolated within different developments.

Al Muth said that this piecemeal mitigation can only work in certain environments. A seaside cliff, for example, may be an independent environment, and may continue existing independently in spite of development upon the land behind it. In the Coachella Valley, though, the small preserves would harden as surrounding developments halted sand from blowing in, and as the old, loose sand blew away. The fringe-toed lizard would face demise because of piecemeal mitigation.

Muth also explained what biologists have learned, through island studies, about the extinction rate of species. "Species go extinct faster on an island than on the mainland. And on islands, species go extinct in proportion to the size of the island. Essentially, a desert preserve surrounded by condominiums, shopping centers, etc., becomes an island. The bigger the island, the fewer species that vanish."

Everyone in the meeting room wanted one, large preserve.

YVONNE P. TEVIS

Developers wanted one so they wouldn't have to chop expensive land out of every development, thus decreasing the size of their projects, and also so the lizards and conservationists would stay in the big, desert preserve and be quiet.

Environmentalists wanted a big preserve to save all desert wildlife. Also, a large preserve could appease a community sick of hearing about the fringe-toed lizard by presenting the cities with something like shared stock—wild desert land. People would use a big park, whereas no one would visit or probably even know about a fragment of habitat set aside here and there.

The USFWS' Kobetich proposed that to arrange the preservation of a large, contiguous area of lizard habitat, a developer could pay a fee corresponding to the number of acres he planned to develop. The Lizard Club would collect these mitigation fees to buy the big preserve, and the USFWS would grant developers an ESA 10(a) permit.

One thing developers dislike as much as a surprise is losing time. They said they didn't want to have to wait around, not developing, until enough money came in to buy a preserve. Collecting the money would be nearly impossible anyway since a mitigation fee couldn't be imposed unless the developer took out a regular request for a building permit as cities and counties require. But developers wouldn't request permits unless they knew they would be able to build immediately.

One way for the developers to get moving fast while the second home market boomed would be for the Lizard Club to agree on the mitigation fee price (which would have to be individually approved by the cities and the county) *before* it presented a conservation plan to the US-FWS. Developers would pay the mitigation fee and build before the USFWS gave out an ESA 10(a) permit. The USFWS would have to shut its eyes to the illegal taking of lizards and trust the Lizard Club to come up with a good conservation plan.

The Lizard Club had to convince the USFWS to go along with the idea, and then plot the "Coachella Valley

64

Fringe-toed Lizard Habitat Conservation Plan," which would concretize the strategy to buy a big, desert preserve.

"I thought, 'The whole thing's insane,'" Paul Selzer said. "I'd already done a lot of research. The area they wanted had 900 different owners. The prospect of organizing this plan was overwhelming. I said, 'It's a federal law, let the feds handle it.'"

Gail Kobetich said that, at first, the developers proposed that the BLM trade federal land in other parts of the country with Coachella Valley lizard habitat. "The BLM said no way."

Leslie Cone, Indio Area Manager for the BLM, said, "A lot of misinterpretations have gone around that the BLM didn't want to cooperate. The BLM was against a preserve being totally a federal responsibility because it wasn't totally a federal problem. The lizard is a state endangered species too.

"By then, conservationists and developers had come up with a huge price tag for the lizard habitat. We said we'd only help with a preserve if everyone, not just the BLM, contributed."

Since the BLM wouldn't take on the whole land trade, Sunrise Company suggested that the USFWS purchase the preserve by requesting funds from the United States Land and Water Conservation Fund. The USFWS replied that never before had it received for one project as many millions of dollars as the preserve would cost, and Congress would not appropriate the agency that much now.

Bill Bone decided to charge ahead, and he said Sunrise Company would organize and complete a huge, combined quest for a preserve in exchange for being allowed to continue with planned developments not in the proposed preserve. The CDFG didn't trust him. After all, while Sunrise developments rolled smoothly across the lizard's habitat, the company could put off organizing the acquisition of the preserve.

Dr. Mayhew had faith in Bill Bone's offer. He believed the Sunrise president wanted to legally free his company from the ESA strictures. Dr. Mayhew figured Sunrise possessed both the motivation and the political pull to create a preserve, while few other local development companies did. As a biologist, he testified at a Riverside County Planning Department meeting in favor of Sunrise's Palm Valley project, and the board approved the building permit.

"I've never seen any development go up so fast!" Dr. Mayhew chuckled. "The very next day they were digging. I guess they were scared the permit would get taken away." Said Selzer, "Through the Lizard Club meetings we raised the level of trust. That we've all proven worthy is the great thing about this project."

Dr. Mayhew and Bill Bone broke the tradition of distrust between conservationists and developers. Next, they faced raising money, dealing with numerous landowners, changing the public perception of a scorned beast, and coordinating bureaucrats, builders, and biologists.

Three times since 1979, Dr. Mayhew had tried to persuade California Nature Conservancy to buy a desert preserve. Buying wild places constitutes the business of Nature Conservancy. The national Nature Conservancy owns preserves in nearly every state. The California chapter alone manages thirty-three.

To conduct this business, the conservationists running Nature Conservancy have shrugged off the stereotype of the bearded and sandaled conservationist ranting before a pinstriped Congress about nature and beauty. American business, the traditional enemy of environmental conservation, does not trust bearded and sandaled people. Since American business influences conservation laws, and wields power through manipulating money, Nature Conservancy strives to gain the trust of American business.

Nature Conservancy conservationists wear suits, because they *are* businessmen—executives of major banks and industries who see to it that Nature Conservancy business flows as any business would—hard bargains driven, bills paid, deadlines met. Other businessmen respect the management of Nature Conservancy.[64]

Also, Nature Conservancy keeps on the friendly side of American business by completely avoiding political brawls. The lobbying of conservationists only triggers the enmity of businessmen.

Nature Conservancy conducts its work based on the American maxim that if you want something, buy it. It's yours and then you won't have to rely upon shaky environmental laws that Congress always seems to disembowel during economic turmoil.

A computer determines which endangered lands in the United States Nature Conservancy will buy first. The environment must be unlike preserves Nature Conservancy already owns, and must display a grand variety of wildlife and scenery. The more pressing looms destruction, the higher priority the land wins. But a computer never sees the wild lands. Someone who does must finally say, "Yes, we will buy that."

In his San Francisco office Steve McCormick, California Field Representative of Nature Conservancy, organizes fund-raising drives, and strategies of acquiring land. From the partition walls hang huge photographs of barren, dramatic landscapes—canyons, parched deserts, and mountain ranges marking the horizon. Although comfortable, the office with its makeshift partitions and cardboard boxes stacked around seems as low budget as possible without offending rich donors.

Much of Nature Conservancy funds spring from outright donations of money. Other times, a person or a business donates a building or a piece of land, and takes a tax break. Nature Conservancy does not hold onto houses or empty city lots, but sells the property and buys a preserve.

Bill Bone chartered a plane to fly Paul Selzer and Dr. Mayhew to San Francisco. Dr. Mayhew credits Sunrise for possessing the business clout to entice Nature Conservancy. Yet, in spite of Sunrise, Nature Conservancy involved itself with the Lizard Club through a lucky coincidence of circumstance.

The Richard King Mellon Foundation decided to hold its convention in Palm Springs, perhaps because one of the Mellons had just moved to the desert. The Mellon

Foundation has supported Nature Conservancy extensively. It had even recently set up a $25-million-fund to preserve the nation's wetlands.

To thank the Mellon trustees for their support, Nature Conservancy wanted to throw them a conservationists' party. Steve McCormick asked Dr. Mayhew about a good spot to take the trustees for lunch. Dr. Mayhew told Al Muth to give an inspection tour of Thousand Palms Oasis to Pete Seligman, director of the California Nature Conservancy Field Office, and Steve Johnson, director of stewardship. "Serendipity," Al Muth reflected. The trio banged on Paul Wilhelm's oasis house door. He took them around the oasis and talked of the oasis' peacefulness, of the Indians and of the wild animals. They had a great time, and were eager to show the Mellon trustees the oasis too.

Nature Conservancy conducted the trustees' desert junket. The trustees could have been displeased: They had been thinking of wetlands, and yet they found themselves bumping around a desert in a tour bus with view windows revealing all the barren stretch. But that spring day shone like those that have made Palm Springs a world-famous resort—days of sunshine and crystal blue sky.

The Mellon trustees got out at a very wild palm grove in Thousand Palms. They walked into shade where trees lean, dry fronds rustle, and the arid, cool smell of an old place accentuates the silence.

The trustees climbed back into their bus and drove down the road and turned into another grove, one of water. There, in Thousand Palms Oasis, they wandered delighted to the reedy pond that glimmers in the desert sunlight. They explored the thick shade below those odd, fantastic palms with fat heads from which bird song sounds like water. They followed old stone steps to a clearing.

Paul Wilhelm said, "They were met in the clearing by a whole catered lunch. Tables were covered with white cloths and festooned with baskets of fruit and orchids. They lunched on beautiful prime rib sandwiches and very cold white wine. They enjoyed themselves thoroughly."

That afternoon, they left the desert.

In cold New York they remembered the oasis, and feasting beneath palms in the clear, desert air.

Nature Conservancy later sent them a grant proposal suggesting that a desert with palm oases occasionally flooded from sudden rainstorms makes a wetlands.

The Mellon trustees remembered Thousand Palms, and from the Foundation's $25 million wetlands preservation challenge, granted Nature Conservancy an interest free loan of $2 million to purchase the oasis and preserve it as they had seen it that vivid day.

When Pete Seligman and Steve Johnson of Nature Conservancy went to the oasis before the Mellons' tour, they also explored the desert to see if Nature Conservancy would want to create a Coachella Valley preserve. Driving past the yellow, wood frames of construction sites convinced them that wild desert vistas soon would vanish. Those desert vistas—rolling sand dunes, silt-striped arroyos, mesquite fields, the desert of roadrunners, coyotes, shrikes, snakes, and fringe-toed lizards—opened up a wilderness which no other Nature Conservancy preserve yet protected.

The Nature Conservancy executives had come seeking a preserve at a good time. Nature Conservancy's grant proposal to the Mellon Foundation pointed out why Nature Conservancy felt sure of success if it took on the task, with the Foundation's help, of preserve acquisition: Local developers and conservationists, previously bitter opponents, had begun compromising for a preserve they'd safeguard together. Sunrise Company had determined to break the development deadlock and would lend Nature Conservancy local political muscle. Nature Conservancy itself wouldn't be trying to squeeze water from a stone as it sought offerings of money and time. The wealth of the valley gleamed from golf courses, boutiques, restaurants, and expensive cars. The Mellon trustees saw this wealth the day they visited the desert.

Nature Conservancy wanted the Mellon Foundation to protect, among the green and silver resort landscape, a desert wilderness. Steve McCormick said the Mellon Foundation seeks challenges, and so Nature Conservancy made an extraordinary request—a grant to save a desert wetlands. Nature Conservancy never considered merely setting aside sand fields for the lizard, but instead wanted to create an unusual preserve through purchasing wetlands, usually arid, sometimes flooded, and which included palm oases. Gail Kobetich of the USFWS said, "It takes a highly sophisticated mind to see a desert that floods out every few years as a wetlands."

This noncomformist view of wetlands captivated the Mellon Foundation, which in turn established a $2 million rollover loan out of its wetlands preservation fund. With the $2 million, Nature Conservancy would buy Thousand Palms Oasis and then raise money locally to repay the Mellon Foundation. Actually, the repayment would be rolled-over back to Nature Conservancy as a preserve management fund.

Steve McCormick called Gail Kobetich, who he knew had been roughing out a conservation plan to resolve the clash between federal law, and people pushing country clubs. Kobetich told McCormick the USFWS wanted a solution, and not to prolong a struggle for territory during which the threatened lizard's numbers would shrink further.

Ready to gamble on a federal agency's willingness to accept an unusual answer, McCormick met with the Lizard Club. He explained that Nature Conservancy could help buy a preserve, but that this preserve would have to embrace land beyond lizard habitat to provide blowsand to the dunes, and it would have to include palm oases, the green show of water of Nature Conservancy's Mellon-backed, desert wetlands. The price for the property—$25 million.

The factions within the Lizard Club panicked, each scared it would get handed the most responsibility for saving the lizard. They hollered, ready to dump the project. But McCormick then suggested "Let's spread the burden." The BLM, CDFG, USFWS, developers, and Nature Conservancy would each have to raise a block of money.

Sunrise attorney Selzer said later, "Success is due to no one group having to shoulder a burden greater than what they felt comfortable with. This is a win-win situation. It costs everyone, but in turn costs each group less when it figures out what it will get back: developers have no more hassle; researchers can study desert biology...."

The Lizard Club, now including Nature Conservancy, spread the burden like this: Nature Conservancy, through the Mellon Foundation, contributed $2 million for the oasis area. The CDFG had already bought ten acres in the preserve[65] and planned to spend around $1 million. The BLM offered a $5 million land trade with Nature Conservancy. The club figured the developers' mitigation fees would eventually total $7 million. The rest of the buying price Nature Conservancy and Sunrise Company would wrangle from local people and from federal land acquisition funds. (Eventually, they counted on $2 million from the federal Land and Water Conservation Fund.)

A few local people resented Nature Conservancy's moving in to boost local conservation, not realizing that Nature Conservancy could take the drive for a preserve beyond the local level. Nature Conservancy, recognized nationwide for business foxiness, social finesse, and skill in marketing conservationist, aesthetic ideas, could appeal to a worldly crowd wintering in Palm Springs and which doesn't often get interested in local concerns. Also, since Nature Conservancy had no connections with the bitterness in the valley over the fringe-toed lizard, it could assuage angry residents fed up with lizards as no local conservationists could.

In turn, a preserve in the Coachella Valley offered Nature Conservancy another opportunity to spread its name to influential, wealthy people who'd not bothered often with conservationists.

Nature Conservancy only expected to succeed at this complicated project of raising $25 million from government agencies (which are noted for not cooperating together) and from developers. Said Grant Werschkull, who Nature Conservancy appointed to direct the project locally, "We wanted to take it on *because* it would be tough. Anytime you can succeed at something bigger, more expensive, more ambitious than what you've done before, the prestige you gain will make the challenge more than worthwhile, and then you can shoot for even bigger projects."

Nature Conservancy then commenced its strategy to win over the hearts of desert visitors and desert dwellers to the desert.

Dr. Charles Allard, who'd purchased 19,000 acres, much of it critical habit, including Thousand Palms Oasis and a big hunk of the proposed preserve, wanted free of the tangle of bureaucrats, developers, conservationists, the ESA and lizards. His realtor, George Whitney, pounded a "For Sale" sign on the property.

Thousand Palms Oasis went into escrow to developers.

Al Muth got scared. If Nature Conservancy lost the wetlands oasis, and hence the Mellon support, it would back out of the Lizard Club. Then, the fringe-toed lizard preserve would stalemate. Muth wrote a letter to the prospective buyers explaining "critical habitat" and the restrictions imposed by the ESA. He informed them of the area's floods and the expense of flood control.

The escrow fell through.

Furious, Dr. Allard had his lawyers send Muth a letter that scared the biologist enough to rush to his own lawyer. "I didn't hear more about it because they had no legal grounds for suit," Muth said. "I only told the truth. Otherwise, they'd have had me in court."

George Whitney scowled at the name Al Muth. "That oasis area alone would have sold for $4 million. My guy just wanted to sell his property and make a clean profit. He originally paid Dart $13 million *cash* for his whole property around Thousand Palms. When someone pays money like that, I say he's *entitled* to a profit."

Apparently, Dr. Allard decided that to strike a deal seemed wiser than to hang onto property the government wanted. He agreed to sell a part of his land—sand dunes, canyons, bluffs, and oases to Nature Conservancy and the government-for $13 million.

Said Whitney, "Dr. Allard is all for a desert preserve, he's just mad the way these environmentalists tried to push him around. What if Dr. Allard had decided *not* to sell, and just wanted to hold onto his land and see what would happen with the ESA and everything? A dozen developers in the valley would be broke right now. Palm Valley—*everything* going on is illegal. Everyone screams Allard's making a killing, as though that's bad. But he hasn't made a red cent. When he bought, interest rates were fifteen percent, so having his $13 million tied up in this land has *cost* him a lot in interest."

Bulldozers. as Whitney said, "take" lizards in direct defiance of the ESA. Dr. Mayhew dramatically labelled the understanding between developers and conservationists the "balance of terror." The USFWS was depending on the Lizard Club and local governments to work out the conservation plan and get it approved by the cities and the county. As long as the developers willingly pay the mitigation fees, the USFWS ignores illegal development. If the developers don't pay, the USFWS can bring them to face the wrath of the law. In turn, developers collar the USFWS and conservationists into not wielding the force of the ESA because the environmentalists realize that their only hope for establishing a preserve lies in cooperating with developers. Environmentalists need both developers' money and their agreement to not build illegally on the desert planned for the preserve.

Developers have been paying mitigation fees before federal acceptance of the conservation plan either because of the promise of no future hassles and their own spirit to cooperate, or else because of persuasion from Bill Bone, and city halls who can give or withhold building permits. They have not paid mitigation fees from fear of the USFWS. Not only could developers have banded together and sued the living daylights out of the ESA, but also it is cheaper for anyone developing seventeen acres or more to break the law and pay the $10,000 dollar fine per violation than to "cooperate" with a $600 per acre mitigation fee. Seventeen acres times $600 dollars equals $10,200.

Not all developers cooperated. Late in 1983, the R. A. Glass Company graded and planted with grapes 220 acres of the proposed preserve.[66] Two weeks later, the county imposed a moratorium on agriculture. Al Muth said, "They did it to beat the moratorium because they knew that within two weeks they'd have to apply to the county for a permit, which they'd never get." He said he abhored the lack of morality behind the grading.

Furthermore, the company had violated the ESA by "taking" lizards, and Muth demanded prosecution. "If I'd known then what would happen, I would have kept quiet." The USFWS overlooked the vineyards on the preserve. C-VERF got the statewide Desert Protective Council to threaten to sue the USFWS under section eleven of the ESA. The government took action against R. A. Glass.

"It was not a vigorous prosecution," said Muth. "For the two instances of grading, the government could have demanded something like $20,000, instead they opted for a softer punishment."

The punishment could have encouraged developers to break the law. If R. A. Glass had waited to pay the mitigation fee on its 220 acres, the company would have paid $132,000. Instead, it paid a fine of $10,000.

"I don't know why, but no one else followed that example. I never would have said anything if I'd known the USFWS wouldn't be tough," Muth later admitted.

Muth said that Nature Conservancy might take out a development option on some of the fields, partly for experiments in environmental regeneration and also because the grapes block the wind-driven sand from 160 acres of habitat. R. A. Glass will have the area for twenty years, after which Nature Conservancy buys the property at the price set in the option. By then, R. A. Glass has made its profit.

"We'll see how the wind-shading has affected the preserve," commented Muth.

While the Lizard Club chewed over the conservation plan and Al Muth watchdogged the property, Nature Conservancy and Sunrise Company chased $10 million in federal funds. The USFWS has a Land and Water Conservation Fund with which it buys habitat. The United States Congress must approve the spending.

Steve McCormick was visiting Washington, D.C., on other Nature Conservancy business when friends there mentioned that that day was the last in which anyone could add on Land and Water Conservation Fund requests for that year. He rushed to sign up the Coachella Valley Preserve.

Later, when Congress considered Land and Water Conservation Fund requests, Al McCandless, the congressman for the valley, pushed for the appropriations to the Coachella Valley Preserve. "Our friend Al," said Al Muth "isn't noted for his support of conservation issues. That's where we couldn't have moved without Sunrise. Paul Selzer really worked on this job."

"It was a job I was paid to do, and paid well," Paul Selzer smiled. "As far as getting McCandless to introduce the bill....Well, the law of the ESA says there's no taking of the endangered species. It takes maybe ten minutes to explain the law and to explain that we've got to get the money or the ESA will block housing, which hurts jobs, which affects votes. After all, the fund was already *there*; he didn't have to go get the money.

"Actually," the Sunrise lawyer continued, "Nature Conservancy has a lot of influence in Washington. They say they don't lobby, but that's crazy, how else do they get things done? Nature Conservancy is involved with lots of influential people who can pressure in small, effective ways."

He said that Nature Conservancy knew exactly who to see in Washington so that the appropriations would be approved. It's Nature Conservancy's job.

"I went with them to the budget committee, to the Senate this and the Senate that," said Selzer, who relished his political tour of D.C. "At first, we were just going to ask for $5 million. 'Gee,' I said, 'let's just go for it. Ask for $10 million.' We got it.[67] We were able to say to Congress, 'Look, we've got businessmen and congressmen and scientists behind this.' The old boys were thrilled. That's the way the country's supposed to work."

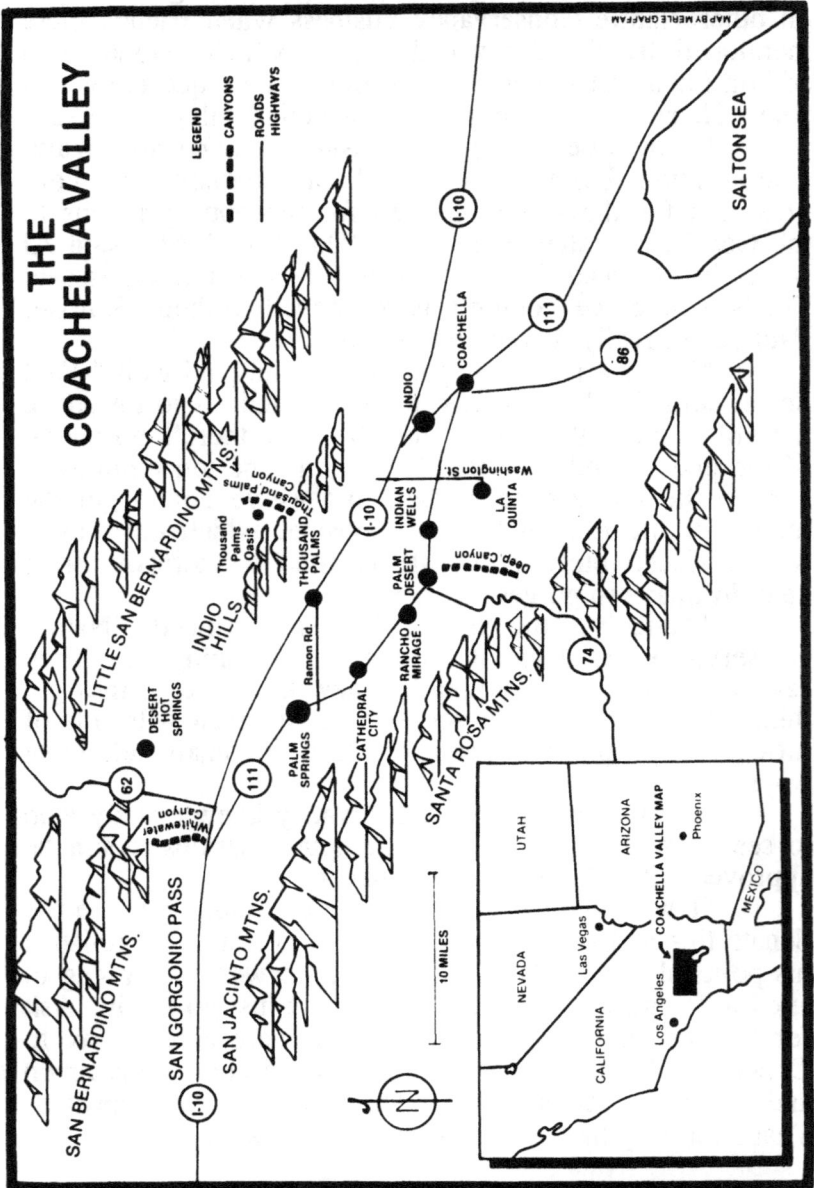

Map I: The Coachella Valley

VI.

A DESERT WETLANDS

Ever since college, Grant Werschkull had wanted to work for Nature Conservancy because, he said, "it achieves results." In college he had engrossed himself in environmental projects, enjoying the politics and psychology of conservation. He became familiar with businessmen and politicians.

"I learned that big business and industry aren't always the bad guy," Werschkull said. "The conservation group I worked with actually had to sometimes put the heat on the Forest Service, and the Fish and Game instead."

He met a mentor, a doctoral student finishing his thesis on old-growth environments. "He really took me out of the 'tennis-shoe-and-jeans' stage. He could say 'Write this letter to this company president,' or, 'We have to go meet the governor.'"

Because of Werschkull's own record of conservation work, including a post-graduate degree, he got a job with Nature Conservancy, and was sent to the Coachella Valley to orchestrate the preserve acquisition project, and to round up $2 million to repay the Mellon loan. Paul Selzer's law firm, Best, Best, and Krieger, settled him in an office "with a door and everything," and for the first months Sunrise Company put him up in its lavish Monterey Country Club for a low rent.

Werschkull knew something of the local discord, but, never having lived in the valley before, couldn't comprehend its bitter depths. Like Al Muth, he plunged in when some long-time residents would have said that to mediate successfully between Coachella Valley factions called for a miracle.

Entering the drama as a neutral character, Werschkull became the mollifier with a wide smile.

"I can't get argumentive and offend people, but must be a diplomat," he said. "One night, though, I was with friends of mine, a couple. She was making a lot of what I thought were assumptions, going on about how Palm Springs is such a cultural area with a variety of people living here. Irritated, I said, 'I think this is a cultural desert....' And then I caught myself up. We seem to have accepted 'desert' to mean a wasteland."

Werschkull does not see the desert as barren. He sees the bare mountains, the crust of sand, and savors the brown and white space. The first time that I went to his office to interview him he said, "Let's go where we can see the mountains." While we talked, parked on an empty street, storm clouds moved around the mountains, hiding the light, and mixing it, and dropping it in splashes upon the hills.

He explained how Nature Conservancy tackles being new in town with a big project to sell to people who appear more interested in the Bob Hope Desert Golf Classic than in protecting desert.

"Steve McCormick," said Werschkull, "has had lots of experience in making nature acceptable in areas like Palm Springs where the atmosphere is social," areas where residents will respond to an influential business and society name such as David Packard of Hewlett-Packard, and a member of the Board of California Nature Conservancy. David Packard knows a lot about how to enchant urban people with nature. He transformed an old sardine cannery along Monterey's Cannery Row into the gigantic indoor-outdoor Monterey Bay Aquarium, with the upper crust of Carmel and Monterey doing volunteer work. At night, they rent the aquarium for exclusive parties.

Packard wrote a letter for Nature Conservancy and sent it to former President Gerald Ford and to Walter Annenberg, former U.S. ambassador to England, both Coachella Valley residents. He said he hoped they would join him in supporting this remarkable plan to create a preserve that would stand out from everything else in the valley.

Both men joined Nature Conservancy's desert preserve drive.

Even with this introduction to the valley, Werschkull had to contend with the lizard. "Fringe-toed" had rung an ugly chord for a long time, and for centuries of human history, "desert," the habitat of the fringe-toed lizard, has connoted a dry, inhospitable place. Once, a Nature Conservancy docent showed a ladies' club the proposed preserve. "Where are we going?" asked the president beforehand. "The oasis," the docent replied, "otherwise you'd be hiking across the sand, and into the canyons." Summer temperatures had descended upon the valley. "We don't want to do *that*," the president said, "take us where it's pretty."

"Pretty" means water and trees. Los Angelinos, or Texans, or Canadians pay as much as $500,000 for a condominium. From the public road they drive through the club portals where two fountains gush into pools on either side. Lakes sparkle on the golf courses. Vacationers crowd around Olympic-sized, hotel swimming pools. Shopping plazas refresh customers with the sound of fountains. Water, water, everywhere in a desert valley.

Fighting water with water, and $2 million, evolved as Nature Conservancy's strategy to save a huge, habitat-varied preserve. When Nature Conservancy arrived in the valley, it bought, and protected from development, Thousand Palms Oasis. The trees and water received lots of publicity. People began to think of the attractive oasis as the symbol of the whole preserve. At the heart of the preserve lies the oasis, surrounded by rough hills and dunes—the space of raw desert. Thousand Palms Oasis has, like a heart, pumped blood of social acceptance and money to the failing space.

The Palms Springs social season begins in the fall when winter residents throng to the warm desert. Nature Conservancy joined the 1983 party circuit, and with the Palm Springs Desert Museum hosted a black tie dinner for the debut of an expensively produced film about the preserve. The summer before, in the relentless heat, the film crew had put together a montage covering the scope of the battle to preserve desert: shots of the little lizard doing push ups in the sand; a coyote trotting to the beat of background drums; then a roadrunner scampering. Dr. Mayhew tells about his long dream. Citizens in front of shops lament the horrendous growth of the valley. The camera whizzes along fast-food restaurants and motorcycle stores, and irrigated vineyards, past floodwaters that pour through the washes and city streets, and whizzes out of the valley to a Riverside County Board of Supervisors' meeting where Dr. Mayhew, Paul Selzer, and developers converse amiably. Then follow shots of the oasis: Steve McCormick of Nature Conservancy, encircled by palms, describes the Conservancy's goals. Water trickles, sunbeams angle through the dark forest of skirted palms.

The film traveled from parties to club meetings and to business meetings.

To stir zeal for a preserve, even in the living rooms of Coachella Valley people, Nature Conservancy mailed out a portfolio telling of Nature Conservancy's ambition to save part of that desert of varied habitats that lies beyond the neighborhoods where the prospective donors live, and telling how through combining efforts with developers, government, and the Richard King Mellon Foundation, Nature Conservancy would succeed. In the portfolio, color photographs vivify desert wildlife, while the cover shows the heart of a desert wetlands, a dark green oasis rising where the brown Indio Hills join the white sand.

The daily job for Nature Conservancy's Palm Springs staff, Grant Werschkull and his assistant Ann Gowman, meant assuaging irate landowners, holding press conferences, smoothing rifts between developers and conservationists, arranging land deals, procuring support from politicians, and incessantly talking to people to build up the splendor of a preserve.

Rather than tire the public with too many publicity parties, the Conservancy waited until that May to give a California Round-up. The invitations read: "President Gerald Ford and Nature Conservancy invite you and your guest for a Breakfast in the Oasis." The required dress—old California style, perhaps to recall the stagecoach watering hole era of the oasis. A ritzy caterer, The Butler Did It, served breakfast. Tables adorned with cacti in Mexican pots reposed in the shade. Chefs cooked eggs to any request—with no waiting in line. Waiters wearing white gloves glided soft-footed as coyotes amongst the palms. Signs on the tables proclaimed, "The Coachella Valley Preserve has been identified as the last possible site where this dynamic desert heritage can be protected." Former President Gerald Ford spoke. "I believe that it's a vitally important project of the great state of California and the country as a whole."[68]

Dr. Mayhew ate eggs in the oasis too, but did not speak. His neighbor at the table joked, "Why are you so unobtrusive?" The biologist responded jovially, "I'm only water boy now." Dr. Mayhew had gladly stepped back as the biologist branded by the lizard, and let Nature Conservancy take charge of persuading monied people in the Coachella Valley to believe in preserving desert.

Money, Werschkull said, can realize visions. His job means figuring out what visions people want to believe in, and then inspiring them enough to donate. He calls it the science of philanthropy. Nature Conservancy's entice-ment of donors began with disassociating itself from the controversial lizard past. I said to Grant Werschkull that Nature Conservancy nearly had covered-up the lizard. He exploded with an uncharacteristically vehement "No! Ab-solutely not. The fringe-toed lizard is extremely important. Without it on the endangered species list there could be no preserve."

"Yes, but we don't hear about the lizard much any more. We don't daily read hate letters in the papers." In-stead, the romance of an oasis in the desert imbues the whole preserve.

"I'm glad to hear that," he said. "Because that's what I've tried to do."

(He remarked that no one's forgotten the fringe-toed lizard. Hiking one day in the mountains above Palm Springs, he slipped and fell thirty-five feet. Afterwards, with his arm in plaster he walked into a developer's office.

"I didn't even think they knew who I was, but when the secretary saw my cast she said, 'All this for a lizard?'")

To the cooling oasis Werschkull lures potential donors. "Biologists kind of like going out in the sand," he said, "but it's hot, so not everyone's going to get into that. One day I took people from the Gallo Winery to an overlook point in the mountains so they could see the whole preserve. They thought the view was incredible. It's nice to have a place to be able to go like the oasis where we could have orange juice and croissants and talk."

Along with enticing donors, Nature Conservancy seeks volunteers. From the day they arrived in the Coachella Valley, Werschkull and his assistant began storing in a docent file names of local people who indicated interest in the preserve, and who Nature Conservancy could ask to show the film at a club meeting, lead a tour to the oasis, or help with a party. The more names in the file, the better, because then a certain few don't feel pressured or get burned-out, or become provoked with feeling like they are the ones who "do everything." Even if a docent only helps with one activity, he will show more interest and pride in the preserve than if he'd contributed nothing.

Werschkull said a preserve could richen the valley beyond setting aside wilderness. A spirit of pride and co-operation stemming from supporting a valley preserve can link all the cities and clubs even in this winter-tourist community where people live behind walls and the loyalty belongs to the country clubs. He added, "If Nature Conservancy can get other groups to take on the management task, that's great. Then people don't brush off the needs of the preserve assuming everything's taken care of. Interest is constantly rekindled."

For example, the Palm Springs Desert Museum led a tour of forty people from Beverly Hills to the desert wetlands. On other occasions, the Living Desert Reserve conducts tours for Nature Conservancy. A sign at the beginning of the McCallum Trail in the oasis says, "For a deeper introduction to the fascinating world of desert animals and plants visit the Living Desert Reserve, in Palm Desert."

Nature Conservancy likes to involve local groups because these have learned techniques for getting business done in their community. To ignore local wisdom and swoop in with a Master Plan That Works Everywhere would be like tossing encyclopedias in the trash. Instead, Nature Conservancy has asked the Living Desert Reserve staff to write signs describing desert biology along walking paths interlacing the palm groves.

Werschkull found a local artist to design the benches that would periodically relieve walkers. "I spent one whole day with the artist, going over sketches which I can then show to donors—who might like a plaque on a bench for $50,000."

Such plaques represent the science of philanthropy as it practices alchemy, turning what once appeared dull rock into a jewel, re-creating desert as an appealing wilderness and a worthy charity. For the McCallum Desert Foundation, preserving the McCallum name upon the natural desert was reason to give $300,000 towards the preserve.[69]

More than one hundred years ago, Judge John Mc-Callum was brought by an Indian friend to the desert, where he saw mountains, dunes, and softly colored hills. He envisioned his son cured of tuberculosis, his family living salubriously in a valley of prosperous ranches which would send throughout the country the earliest ripe fruits of the season.[70] He then moved his family to the desert.

The judge dreamed that Palm Springs would grow into a paradise. He bought vast acres of land from the Southern Pacific Railroad, and advertised for settlers. Newspaper ads shouted, "No frost, no heavy winds, no fog. The home of the banana, date, and orange."[71] To irrigate vineyards, Judge McCallum had workers build the Whitewater Ditch, a narrow, stone-lined canal harnessing the stream flowing from Whitewater Canyon.[72]

The judge's five-year-old daughter, Pearl, on moving day rode into Palm Springs on the front of her father's saddle. She lived there the rest of her life. When she was eighty years old, bewailing that she could no longer ride horses across the desert, she said, "That's what's wrong with me, I need the desert."[73]

Pearl McCallum McManus devoted herself to making Palm Springs a resort. She built tennis courts by her pink house on the hillside, and later added a swimming pool where two palms rose in a V beside the water. Finally, she built a clubhouse, and so commenced the first tennis club.[74]

The two faces of Palm Springs had evolved: the solitude of the desert, of space filling a valley to the top of jagged mountains, overflowing into the sky, and also the conviviality of the social life—tennis, swimming and suntanning, clear air filling the lungs, sun and health overfilling the body.

To Pearl McCallum McManus both aspects of Palm Springs produced the goodness of desert life. She backed the California Institute of Technology's mobile, desert-research laboratory—a trailer equipped for biologists to use to explore the desert. She commissioned a Mexican architect to design the large, stone fountain on the entrance lawn of the Palm Springs Municipal Airport.[75] As people come and leave, they see fruitfulness, a fountain in the desert.

The McCallum Desert Foundation, which Pearl McCallum McManus had established in her will,[76] was an obvious group for Nature Conservancy to appeal to for donations. Judge John McCallum had pictured a village in the desert. The Foundation could carry on the McCallum vision, but instead of building the city, the McCallum name could preserve something rough and rare within an expanding resort.

However, any foundation receives many requests from reserves, museums, hospitals, schools, churches, temples, and foundations for the handicapped. Said Werschkull, "We knew we had to give the McCallum Foundation something different back." Like the ancient Greeks, most everyone craves immortality, but the desert, merely "preserved," doesn't recall people to posterity. Coins swallowed by a common donations hat lose their individual face. So, in exchange for $300,000 from the McCallum Desert Foundation, Nature Conservancy mounted a plaque on a water-rolled boulder smack in the main clearing of Thousand Palms Oasis. It reads:

> "The wild and open spaces of the desert were home to the McCallum Family who settled in the Coachella Valley in 1884. Preservation of the pristine desert lands protected in the Coachella Valley was made possible by a generous gift from the McCallum Desert Foundation, established by Pearl McManus, in memory of her father, John Guthrie McCallum, desert Pioneer."

The McCallum Trail winds through the oasis and out along a dry stream bed towards another palm stand renamed the McCallum Grove.

Nature Conservancy commemorates other people in Thousand Palms Oasis where Paul Wilhelm had built a house out of boards of palm. The rooms display Indian artifacts, and literature about the preserve. On one wall hangs a list of the original big donors—the Palm House Club. Each member keeps a key to the house.

To find new donors, Werschkull would go to a government meeting and talk to someone, and make another contact through that person, and get invited to another cocktail party, and meet another person who would ask him to speak in front of another group. He followed through on the introductions, meeting people to introduce the preserve to friends when he wasn't around.

In this way he tapped the Desert Riders, a club which brings together a group of avid horsemen, many who have lived in the valley a long time. Once, the night before the scheduled ride would cross the preserve, Werschkull called up Jean Hahn (wife of the developer of the mall which had so angered Al Muth).

"It's only natural they'd all talk about the preserve," he said. "I wanted to chat with Jean, who's really behind the preserve, and subtly remind her of the preserve's aims without telling her to ride out there and get us some money."

He told her about the system of equestrian trails that would network the preserve and how the big, riding area would inspire people to build small ranches around the periphery of the preserve.

Spurred by Jean Hahn's enthusiasm, talk arose easily during the ride about how the preserve ensures open, desert floor for riding. Afterwards, several Desert Riders made donations to Nature Conservancy for the Coachella Valley Preserve.

The strategy to build upon local support is simple: Nature Conservancy staff talks up a preserve to people who don't work for Nature Conservancy, yet who later, in a sincere and natural way, spread their own enthusiasm to friends. Subconsciously, they draw out those facets of a preserve which mean the most to themselves, and most likely to their friends also.

A desert rider who took the ride with Jean Hahn across the preserve later introduced Werschkull to a wealthy resident of Smoke Tree, an area of Palm Springs where many Desert Riders and Palm Springs old-timers (some even remembering the dude ranch days when desert space seemed too huge to vanquish) live. Werschkull took the Smoke Tree resident to the oasis.

"She's very old and frail, and in a wheelchair," Werschkull said. "But I drove her right into the oasis and we sat in the shade. I think it meant a lot to her to be able to be there. I'd never met her before and although I knew she *could* donate, I hated to jump for the throat immediately when I'd just met her. We simply talked about the preserve. The man who introduced us hand-delivered my proposal to her so it could be as personal as possible." She gave $50,000 anonymously, perhaps because she wanted other people to enjoy the oasis as she did that day.

Even among environmentalists, disputes arise over how a preserve should be enjoyed. Some want it fenced off like a country club, allowing entrance only to researchers or other professionals who will use the land as an audio-visual aid for education. But what about people who cherish wilderness without studying it—horseback riders, bird-watchers, poets, photographers, and children who get bored having the desert *explained* to them? If a scientific or educational aim becomes the only ticket into the preserve, how will a city child by chance stumble upon the lure of wild places?

Nature Conservancy dangles any carrot to tempt interest in its preserve—historical, educational, scientific or aesthetic, and chooses not to restrict enjoyment of the preserve to scientific use. For the non-scientists, Nature Conservancy held an open house at Thousand Palms Oasis. That April weekend, the summer's first hot spell had descended. I doubted anyone would care to battle the heat. By 9:00 A.M., a crowd had gathered and people kept coming. True, no one walked the twenty-minute McCallum Trail to the McCallum Grove, but they did explore the oasis, talking to docents from the Living Desert Reserve and the Palm Springs Desert Museum who manned information tables or waited in the shade to tell about the palms and the Indians, or were stationed outside the oasis, sweating profusely, fanning themselves desperately, and showing which seeds Indians ate, and how they fashioned arrow shafts from bushes.

Other docents escorted families around the outskirts of the oasis, over dry streams where floods have piled arroyo rocks, and tracks of quail mark the sand. (One man demanded, "Let's see the fringe-toed lizard.") They viewed the parting in the range of Indio Hills—Thousand Palms Canyon—where floodwaters rush to the valley. Docents pointed out the bird that eats the gluey seeds of the mistletoe, which it later regurgitates in little piles on the branches of the mesquite. The seeds stick, and the mistletoe grows. Docents told about the creosote bush that kills all other plants around it, wiping out competitors for water. Then the groups stepped into the oasis shade and drank lemonade.

Besides introducing the public to the desert's natural places, Nature Conservancy had to assuage small property owners who felt no charity towards a lizard. Many had bought their land more than twenty years ago as an investment for retirement.

"I didn't know the fringe-toed lizard was going to have more rights and be more important than people," Deloris Hurley, one such resident, said. "I don't think I knew the word 'ecology' at the time we bought. And I certainly never knew these little lizards could influence people's lives they way they do."[77]

Nature Conservancy's office weekly received calls from property owners wanting to sell at prices the Conservancy couldn't afford. Said Werschkull, "I feel sorry for them because until we collect enough from developers' fees, we can't buy, and the people keep paying taxes without doing anything with their property." For developers, conservationists, the lizard, and for small property owners, the faster a plan to solve all their requirements could be settled, the better.

Nature Conservancy had aimed to have the $2 million to pay back the Mellon loan raised by that April. But as the hot weather comes, people's minds turn to leaving. Nature Conservancy needed still half a million dollars. Perhaps the residents and visitors had finished with giving for that season. Philip Boyd, a local conservationist, said the Palm Springs Desert Museum had planned on raising funds to build another wing, but had first brought in an independent firm to feel out the valley. The consultants said donors had been thoroughly milked and to try next year.

"All that's true, but it's disappointing," Werschkull said at the time. "I'm going to go back to San Francisco for the summer. There's really no point in my staying the whole summer. In the fall we'll be back.

"It'd be so good to say to the Nature Conservancy board that we'd done it. But if we don't get the rest of the money here, we'll get it somewhere else."

VII.

A CONSERVATION PLAN EMERGES

The Lizard Club swelled the conference room inside the Monterey Country Club when it first gathered to break the development block. Referring to subsequent meetings, Gail Kobetich said, "Those massive meetings were really too big to accomplish much quickly, but they were a good way to get information to the public. Also, they kept things calmer. It intimidates most people to be totally unreasonable in front of a big group of peers." Figuring out a conservation plan everyone would accept demanded all parties together speak, listen, and compromise.

Kobetich helped start the communication by ironing out opposition to the conservation plan from a strong, local agency. The Coachella Valley County Water District had opposed a preserve back when Dr. Mayhew was first trying to get the lizard listed.[78] Because the water district owned critical habitat, Lowell Weeks, the district's long-time general manager, attended Lizard Club meetings.

"I'd always heard what a dragon Lowell was," said the Forest Service's Kobetich, "but we got along very well, even reminiscing about adventures along the Colorado River in the Imperial Valley. At the next club meeting, Lowell stunned everyone with an impassioned speech in favor of the preserve. Everyone else fell in place behind him. They thought, 'Well, if Lowell's supporting it, we can too.'"

Because the Lizard Club's size hindered swift exchange of ideas, the group chose a steering committee consisting of one representative each from the Agua Caliente Tribe, the native Indian tribe which owns some critical habitat; the BLM; the CDFG; C-VAG; C-VERF; the Coachella Valley Water District; Sunrise Company; Nature

Conservancy; the USFWS; and Tierra Madre Consultants and Thomas Reid Associates, the two firms which handled technical authorship of the plan.

The steering committee would work out the wishes and complaints of the various groups, and report back to the massive Lizard Club. The steering committee could not circumvent satisfying each member of the Lizard Club because any member's opposition could convince a city to not sign the plan. Nine cities and the County of Riverside wielded control over the local fate of the plan.[79]

Cathedral City, Coachella, Desert Hot Springs, Indian Wells, Indio, La Quinta, Palm Desert, Palm Springs, Rancho Mirage, and the County of Riverside together had to write a joint conservation plan, thus applying for one ESA 10(a) permit. To do this they had to agree on every aspect of the conservation plan before it could be submitted to the USFWS. Therefore, any city, or the county, could balk at mitigation fees, preserve boundaries, or rules for implementing the plan, thus slowing its joint acceptance.

Along with the steering committee, C-VAG offered a forum for public comments and for discussing the good and bad, as far as the cities were concerned, of a conservation plan that would be best for all the desert communities.

To test how the cities and county felt about the plan, C-VAG voted on it several times,[80] but not all nine cities in the Coachella Valley belong to C-VAG. C-VAG's vote held no actual power. Only the individual votes of the nine cities and the county had binding authority.

Long before the plan traveled from city hall to city hall for final approval before going on to the county, from where it would go to the USFWS, the Lizard Club, the steering committee, and C-VAG faced persuading a myriad of interests to agree on one method both to pave the way for valley development and also to save natural desert. Gail Kobetich told them that he would advise them of the ESA conditions the plan had to meet since it had to ultimately satisfy the USFWS.

The changing and deleting of Kobetich's preliminary conservation plan began.

The burden of buying land would be shared. Each of the four owners-to-be of the Coachella Valley Preserve would use a different method to acquire preserve property. The USFWS would seek funds from the U.S. government. The CDFG would tap state money. The BLM would trade lands with Nature Conservancy. (This would work by Nature Conservancy's buying preserve lands with their own funds and trading these purchases for BLM property elsewhere to sell and so recoup Nature Conservancy's original expenditure.) Nature Conservancy would take out three options on large portions owned by Dr. Allard to be exercised at different times.

George Whitney, Dr. Allard's realtor, said, "We fixed the options in an order so Nature Conservancy would have to buy them *all*. Allard sold the most unusuable portion of the property first—the Thousand Palms Oasis area. The area above it is the second most usable. It could have been developed with golf courses. It's got a good view and lots of sun.

"But the most valuable land for Allard is the lizard habitat below, which is prime agriculture land with water and power lines already in, and also access roads. He's made them buy everything else first before he would turn that over. It's the part they had to buy—where there are lizards." (See Map III)

If it turned out that local governments, developers, and conservationists couldn't satisfy the USFWS that the conservation plan would protect the lizard, or if they couldn't even work together to come up with a plan, Dr. Allard didn't want to be left holding only the least valuable part of his property.

The ways and means of buying land from Dr. Allard or acquiring government funds did not stir turmoil at the meetings as did the assigning of developer mitigation fees. No one minds mitigation fees if someone else pays them. The cities of Palm Springs, La Quinta, and Desert Hot Springs managed to cut some areas out of the mitigation fee requirement.[81] Nature Conservancy, busy keeping up its side of the bargain, got scared; other groups seemed to be weasling out of imposing mitigation fees. Grant Werschkull said the areas of the valley most likely to grow soon must fall under the mitigation fee stipulation, otherwise, $7 million from developers would become a pipe dream.[82]

When owners of 200 acres under county jurisdiction together petitioned the county to strike their property from the proposed preserve, claiming that no lizards exist there, Paul Selzer warned that such deletions could prompt the USFWS to deny a permit. "We'll be back to square one. Whatever we've accomplished the last two years will be for naught."[83]

The president of the Thousand Palms Chamber of Commerce, Edward Wasserman, objected to mitigation fees as unfair to the little developer who has lots of acreage to develop, but little ready cash. Big development companies can fairly painlessly afford the mitigation fees, while the fraction which small lot owners must pay of $600 dollars to the acre doesn't figure out to much. But the small developers face what Wasserman called "legal blackmail" because the entire fee must be paid when a developer first receives a building permit. No one has the choice to spread out payments.

The Lizard Club wanted developers to pay everything at once because it needed money immediately, and the sooner a preserve was established, the sooner developers could unstrap themselves from the ESA. Perhaps over time the Thousand Palms Chamber of Commerce will boast its proximity to the preserve. Never does everyone win. Even so, people complain about the "donation" they're forced to make whenever they take out a building permit. Everyone's for the preserve, they don't want to pay, that's all.

Those who don't mind paying, encounter frustrating regulations. One man planned to build on his lot in a mobile home park. Even though the developer had years earlier subdivided the lots and brought in water and power, each lot owner who buys into the trailer park must obtain a permit to build, and so he pays the mitigation fee. The man's fee figured out to be $75. He went to the Palm Desert City Hall to pay. The staff sent him to the Nature Conservancy office in Palm Springs. Then he had to take the receipt back to the Palm Desert City Hall. Irked, he thought less of the idea of a preserve.

Not much opposition boiled during the Lizard Club meetings against developers of urban projects having to pay mitigation fees. Developers realized they would have to make an exchange with conservationists. Instead, the Lizard Club argued whether ranchers should pay for turning desert into farms. C-VAG had established the mitigation fee for any development, including agriculture, at $600 an acre for thirty years. Ranchers and farmers screamed that the economics of modern farming rendered farming impossible enough without farmers being strangled by a fee they couldn't pay. A rancher from Indio, Robert Melkesian, said, "All of a sudden we're penalized for feeding a nation."[84]

When people think that agriculture and the fringe-toed lizard get along, they muddy their biology. Often they argue, "I've lived here twenty-five years and have seen lots of lizards in the date groves—quite happy-looking, too. Agriculture destroys the habitat, but not the lizard."

Actually, the lizards in the date groves are not the fringe-toed lizard because fringe-toeds cannot exist where other lizards thrive. Even after all the publicity about vital blowsand, a lizard's a lizard to the public. Arguing that the Lizard Club should not lower the fee for agriculture, Al Muth said, "A field of alfalfa is just as destructive to the lizard as a shopping center."

Nevertheless, the fee was switched to $600 per acre of urban development, and to $100 per acre of agricultural development with the remaining $500 to be collected when the owner converts the fields to urban development. In the end, the members of C-VAG, along with the steering committee, agreed to drop farm fields from the mitigation fee requirement as long as they remained in agriculture.[85]

"Besides using political muscle," said Les Cleveland, executive director of C-VAG, "farmers got themselves exempted from the fee because farming is considered the bottom line in land use. If someone is not allowed to at least farm his land, the most basic use, he could sue for inverse condemnation. Also, farmers traditionally have never had to take out a permit to farm. They do if they want to grade beyond a certain cut, but there aren't many restrictions against farming in this country. We didn't want to get taken to the state capital over it, so we excused farmers from the fee."

The proposed conservation plan reads that if a farmer converts his field to urban development, he will at that time pay the $600 per acre fee. However, any fields already in agriculture before August 4, 1982, are excluded from ever paying a fee, even if the field sprouts condominiums. "We chose that date," said Cleveland, "because to be fair you have to pick some date. We have aerial photographs of the valley taken on that day and they show which land was in agriculture and which wasn't."

The mitigation fees must be paid throughout the next thirty years. However, as the proposal stands now, developers only pay the $600 per acre fee up until the point at which $7 million has been collected. After that, the fee drops so that developers who have waited to develop will only pay $100 per acre.[86]

For conservationists, a disturbing omission in the present draft of the conservation plan is the absence of strictures against flood control built in the preserve. The Coachella Valley County Water District may at any time construct flood control works through the preserve. Kobetich said strongly that flood control, which would halt the washing in of new sand by water, would render the preserve "unviable for the lizard."

He said the water district "made a blatant threat to stop the project if any restrictions were made on flood control." Kobetich didn't think the water district could actually prohibit federal acceptance of the conservation plan, but that with its political pull, and using the emotional argument that floods endanger human life, it could block the decision for a long time, during which development would continue to eat up more habitat.

To speed up approval of the conservation plan, Kobetich agreed to not restrict flood control except to write in the conservation plan that the water district should consult with the USFWS before beginning flood control.[87] He

said, "I only hope that when the time comes for considera-
tion of flood control, local people will care strongly enough
about the preserve to not let it happen." He used a
metaphor Dr. Mayhew drew early in the preservation battle
which compares the future Coachella Valley Preserve to
New York City's Central Park. Few New Yorkers would
permit any government agency to chop up Central Park.
Kobetich hopes people will feel as deep a pride for the
Coachella Valley Preserve.

Lowell Weeks, general manager and chief engineer
of the Coachella Valley County Water District, said a lot of
conservationists hypothesize unreasonably about what could
happen to the environment in the future.

"People who don't have to stand up to the public
don't have responsibility," he said. "They can say 'What
if....' I hate what-iffers. The water district showed maps
of where flood control would go, and the maps were approved. I think a solution can be worked out. The water
district *cannot* be a party to destroy a reserve, but in turn,
the reserve has responsibility to people, whose safety comes
first."

The conservation plan's strategy to save the fringe-
toed lizard does not rely solely upon the large, 20-square-
mile Coachella Valley Preserve, but also upon two smaller
areas of habitat—the Willow Hole/Edom Hill reserve and
the Whitewater Floodplain reserve.[88] In the early stages of
drawing preserve boundaries, the BLM had insisted on
more than one preserve to ensure against an epidemic, or
neighborhood cats and dogs, or a flood from wiping out the
lizard population. The BLM already owns acres in Edom
Hill and won't need to acquire much more, while the
Coachella Valley County Water District has dedicated,
though not by title, its holdings in the Whitewater Floodplain as a reserve.[89]

Not all biologists believe that establishing more than
one preserve changes the odds one way or the other for the
lizard's survival. They say an epidemic most likely would
sweep through the entire valley lizard population. They
scoff at the idea that cats and dogs could snag enough
lizards to make any difference, and they point out that

floods have poured across the valley for centuries, yet the fringe-toed lizard has survived. Dr. Mayhew called the two small reserves "technicalities to satisfy the BLM." As long as these reserves don't drain funds meant to acquire the main preserve, most conservationists cheer saving all the habitat possible.

Biologists hesitate to forecast that a prescribed number of acres will protect a certain species because no one can play God. Astute politicians realize that *political* truth achieves great ends. Biologists must sometimes follow the politicians' examples and dare to be positive. Hearing "yes, this acreage will preserve this species," people will back preservation projects.

Since a man-made plan for the environment plays a card game with nature, the steering committee has tried to raise the odds for success by safeguarding land outside the preserve, and so divided all lands within the lizard's *historic* range into three categories: 1) land to be acquired for preserves; 2) land which, if developed, will yield developer mitigation fees; 3) land to be managed and regulated.[90] Public agencies presently own the lands to be managed. "Management" means these agencies will endeavor not to directly harm the lizard, or to block the flow of new loose sand while the agencies construct projects such as flood control, utilities, or windmill fields for electricity production. The regulation of land concerns private property unlikely to be developed soon. Through zoning codes, cities and the county will protect regulated land against heavy urban or agricultural development.[91] Of course, future local governments may choose to re-zone these regulated lands and permit development.

Les Cleveland admitted some people had complained to C-VAG about the progression of valley development without an ESA 10(a) permit. But if the USFWS were punishing developers $10,000 for every violation of the ESA, developers would file so many suits against the ESA that the law might collapse within Congress.

"The cities didn't want to go that route," said Cleveland. "The public criticism and that threat of suits spurred them along to agreeing on the plan and sending it to the USFWS."

Once the USFWS had received the Coachella Valley Fringe-toed Habitat Lizard Conservation Plan, it considered several options while reviewing it. The USFWS could have rejected this draft, or altered it, or approved it outright. Rejection would trigger an explosion of outrage in the Coachella Valley. Since the USFWS had ignored the illegal development, letting the balance of terror control development, and since Gail Kobetich and lawyers for the USFWS had closely advised the steering committee, the USFWS had virtually committed itself to accepting the conservation plan. If the USFWS tried instead to adamantly enforce the "no take" clause of the ESA, developers would challenge in court the ESA—the backbone of the USFWS. Judges, under great pressure, would cripple the ESA and render the USFWS powerless.

The USFWS realized it didn't possess the strength to adamantly enforce the ESA and stop development. The USFWS can fine a Coachella Valley developer for violating the ESA only if the agency can prove lizards existed on the specific site, difficult to do unless prior on-the-spot sighting of the lizard has been recorded. If the USFWS proves the presence of the lizard, the agency may collect the fine, but by then the natural environment has been obliterated, and the endangered species, the reason for invoking the ESA in the first place, further jeopardized.[92]

The USFWS was compelled to grant the permit according to the conditions, or a close modification of these conditions, set up in the Coachella Valley Fringe-toed Lizard Habitat Conservation Plan. If the USFWS denies a permit, not only will the agency be unable to enforce the ESA effectively, but local governments no longer will attempt to control development as they have more or less been doing under the balance of terror. They will abandon both protection of the fringe-toed lizard and creation of a desert preserve in favor of encouraging lucrative development.

Denying the permit would surely depress people who hail the conservation plan as an example to the nation.

Said Cleveland, who was greatly impressed by the cooperation among the various interests, "For too long the idea has prevailed that nothing should get in the way of men. But it isn't right either to say nothing should stand in the way of animals. This plan shows an interrelationship between life forms."

As the USFWS reviews the conservation plan, it has to consider possible ways to save the lizard other than those set up by the steering committee. Most of these it has rejected, explaining why in its summary and review of the Coachella Valley Fringe-toed Lizard Habitat Conservation Plan. For example, acquiring *more* preserves would only mean more funding would have to be raised. Anyway, few undisturbed areas other than that of the proposed Coachella Valley Preserve are large enough to meet the fringe-toed lizard's biological requirements. One way to obtain another preserve in an area now uninhabited by the fringe-toed lizard, or to move the location of the present preserve, would be for the USFWS to create an artificial preserve through continually trucking in sand of the exact grains the fringe-toed needs—like a pre-fab windstorm. Obviously, this alternative would involve expensive and endless chores.[93]

Gail Kobetich said that unless someone delivered some terribly convincing, last minute reason against passage of the conservation plan, the USFWS will approve it, making no modifications, and will grant the cities and county their joint ESA 10(a) permit. After all, not only did the USFWS monitor the conservation plan in its drafting stages, but also the agency wants a preserve as much as any member of the Lizard Club.

After the USFWS grants the ESA 10(a) permit, the road to an established preserve still will lead through difficulties. The actual development rate in the valley may turn more slowly than expected. In that case, the mitigation fees won't raise the forecasted millions. Perhaps the congressional budget committees will not authorize the next $5 million allocations from the Land and Water Conservation Fund. (Kobetich said he sees no problems here since both the House and the Senate have fully supported the appropriations.) In fact, the $5 million, along with all appropriations to federal agencies, already has been cut 4.3 percent because of the Gramm-Rudman bill designed to reduce the federal deficit. Kobetich seemed nonplussed. "We can live with that."

So, with the conservation plan, the residents of the Coachella Valley and the USFWS gamble. To conserve means to gamble, yet no one knows if the conservation plan will definitely reverse man's destructive effects upon the desert. The natural world recognizes no human edicts or petitions, and does not reward an admirable act of human cooperation by automatically seeing to it an endangered species lives. It is this mutability of nature which enthralls some people and aggravates others, and from whence grows the disputes between conservationists and anti-conservationists.

Yet this Coachella Valley Fringe-toed Lizard Habitat Conservation Plan has birthed an amazing coalition of developers and conservationists and will in the most fair and effective way possible, save the federally protected Coachella Valley fringe-toed lizard from extinction. Therefore, even before official granting of a permit, the cities and county operate under the Coachella Valley Fringe-toed Lizard Habitat Conservation Plan, and the USFWS approves.

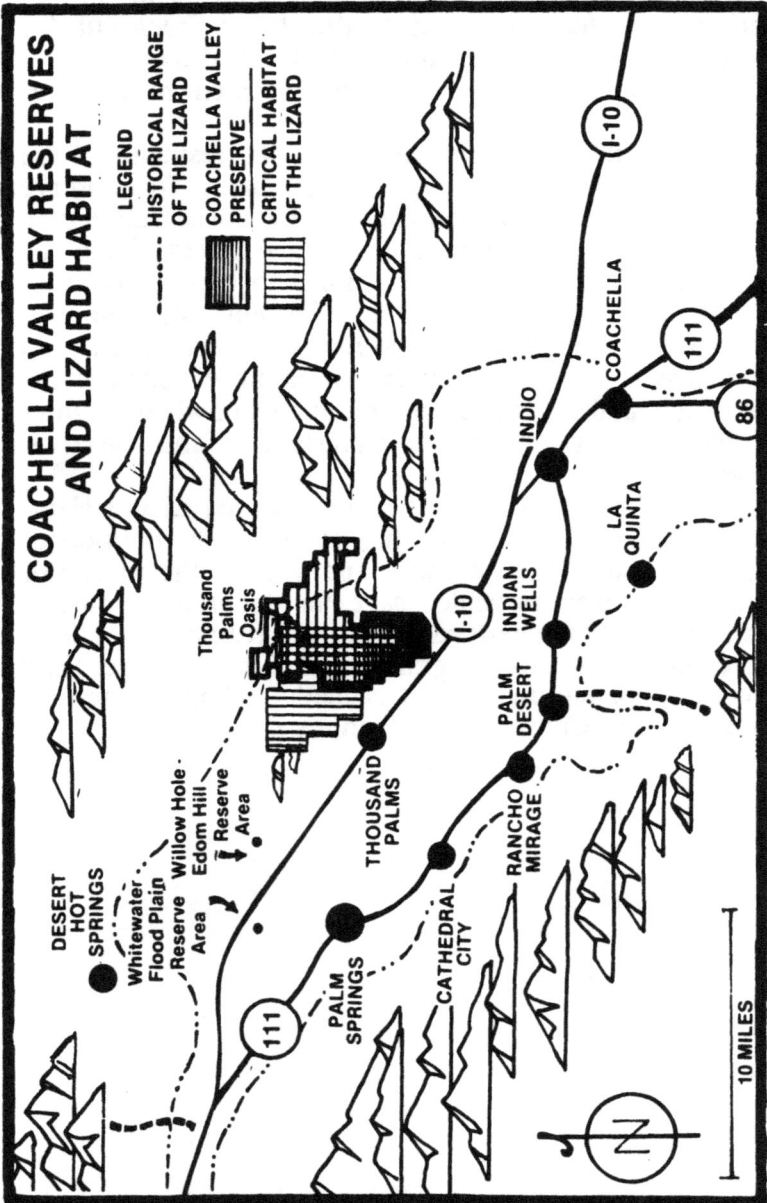

Map II: Coachella Valley Reserve and Lizard Habitat

VIII.

A WILDERNESS PRESERVED

The writer of the *Wall Street Journal*'s front-page column about a tiny lizard confounding development in a world-famous resort described the negotiations between developers and conservationists as being "as ticklish as a Middle East peace plan." The conservation plan's greatest virtue is achieving, the *Journal* wrote, "the preservation of a giant parcel of pristine habitat that will become a kind of Central Park in a desert valley otherwise destined for development."[94]

The article disappointed Paul Selzer. The writer had seemed tickled by the goings-on out west, as though the story were a series of flukes never to reoccur. Selzer joked about the old boys in Congress being thrilled with the democratic triumph of the Coachella Valley solution, but he believes this example of negotiations among enemies can succeed all over the country, and that it must because, he said, "The environment almost always seems to lose; it's the pressure of population."

Although he is the lawyer for a developer, Selzer understands environmental concerns. Although the *Journal* wrote, "The first meeting between Muth and Selzer was like a scorpion convention," Kobetich said the two men worked well together. "Paul Selzer understood the politics of the valley. He hammered from within development circles, and Al Muth hammered on the outside." Both Muth and Dr. Mayhew said Selzer would never have worked so unceasingly for the preserve if the beautifully rough desert hadn't at some time captivated him.

The valley looks very different from when Selzer first arrived, rather reluctantly, at Best, Best and Krieger some twenty years ago, picturing Palm Springs as "all tinfoil over the windows in the summertime." Now the cities are all restaurants, golf courses and air conditioning. So Selzer has brought the desert into his office. Polished rocks hold up books. Large, desert photographs, which Nature Conservancy used in brochures, brighten the walls. One of the cacti centerpieces from the Breakfast in the Oasis sits on a table. That July of 1984 he said that, when the USFWS hands out the ESA 10(a) permit, the old Lizard Club would "all get together for a big party." He grinned.

The first and biggest preserve management project was cleaning up trash dumps. Grant Werschkull talked about getting biker groups to patrol the area for dumping and off-road vehicle trespassing. Together, several kinds of groups organized clean-up days.

Nevertheless, people keep dumping, and dune buggies keep stripping the dunes. "Change," Werschkull commented, "is incremental." He said No Dumping signs don't deter anyone, and he didn't like the idea of policing a park. The public's attitude that a desert's one, big, dumping site only will change over time.

Philip Boyd, a prominent, local conservationist committed to both the Living Desert Reserve and the Deep Canyon Research Center, said he is puzzled how Nature Conservancy plans to control people and the way they treat the desert.

"Other big preserves are set off by themselves, but the Coachella Valley Preserve borders Interstate 10," he said. "From this main access a lot of people will drift in, people who don't understand the desert, and who will throw trash, and trample plants. Of course, water attracts even more people who won't necessarily treat the oasis carefully. So who's going to watch them?"

After *Sunset Magazine* published an article on the preserve in its February 1985 issue,[95] about 100 people visited the oasis every weekend for several weeks. On warm days people put on swimsuits and spread picnics by the water.

Boyd said that the tremendous cost of fencing the twenty square miles of the Coachella Valley Preserve would almost prohibit fencing. After that, staff would have to be hired to collect entrance fees from the public, conduct tours, and keep the place clean.

Werschkull suggested the state park service, experienced in dealing with the public, may eventually take over managing the preserve while Nature Conservancy keeps title to the land.

Worrying how to manage the Coachella Valley Preserve seems a luxury—and means the preserve is no longer a dream.

A few residents still prickle at the word fringe-toed lizard. Realtor George Whitney said, "It's just come out in the last year or so that they got the lizard to use as a catalyst to study an endangered cockroach. But they had no chance with putting a cockroach before the public, so they switched from that to the lizard." Many people still feel that listing a species as endangered is a dirty trick conservationists use to freeze up land and prevent its development.

Dr. Mayhew said that, as he acquires reserves for the university, he uses the political tool of endangered species every chance he gets. But if the word endangered rings like "Wolf!," public reaction will dull. People will get tired of thinking about the earth's species.

People opposing the construction of a water slide park in Palm Springs argued the park would infringe upon the lizard's critical habitat.

"How much taxes did the fringe-toed lizard pay last year?" countered an attorney.[96]

After the Riverside County Board of Supervisors approved the project, Corky Larson, a supervisor for the valley, commented, "We've made a major commitment to the lizard, but we have to say enough is enough."[97]

Public opposition to listing any other valley species would be overwhelming, Al Muth said. "Politically, a desert floor animal stands no chance. The only possibility would be animals in the mountains or the foothills. So if we haven't biologically covered everything in the valley, we can write it off."

Theoretically, the Endangered Species Act aims not to collect a growing list of names, but to preserve a habitat and repopulate a species so it can be taken off the list, no longer in danger. In the early days of the fight for the Coachella Valley preserve, conservationists proposed that if cities set up a preserve, the lizard could be delisted, and valley development could race off unhindered. Muth said delisting the lizard won't happen for a long time.

"What we've done is guess," he said. "We *think* we are saving the lizard, but we want to know if we've guessed right. Anyway, where would the pressure come from? Biologists wouldn't. Developers don't need to because of the conservation plan."

Now, other conservation fights rage. Natural desert beauty has won no permanent popularity. Enough is enough, people say. The Rancho Mirage city council never sued the federal government. Instead, it embroiled itself in a battle over whether or not to permit a resort on the Rancho Mirage mountains. Some residents argued a huge hotel would make the city rich with increased revenue and prestige. Others declared it would impoverish the city. Traffic would clog the small streets, and city funds would dwindle as the city added services to the area, but that above all, the hillside resort—trees, cars, buildings, and lighting at night—would steal the barren wildness from the mountains.

Not lushness, but an emptiness where the soul hits bottom, weaves the magic of the desert. As Grant Werschkull closed the Palm Springs Nature Conservancy office to return to San Francisco, he said he was glad to be going back to big trees, "but in the early morning I look at these mountains. They're so close, like nothing else. I'm going to miss that."

This tale of the Coachella Valley Preserve offers an example for conservationists of how to work with developers rather than alienate them. Development doesn't defy the U.S. Constitution. Developers merely make their living, as do politicians, school teachers, shop owners, and biologists. People need houses and schools and supermarkets. They also might as well enjoy their lives by playing golf, reading in a community library, or hiking through canyons. In the Coachella Valley, developers and conservationists have acceded to one another's rights to live life differently—and together.

An array of characters mixed with twists of fate brought about this accession.

Gail Kobetich said that only through a fluke did he become a prominent figure in developing the conservation plan. He had selected a herpetologist to take over drafting a conservation plan, but then the government put a freeze on hiring for federal agencies. So Kobetich, the only herpetologist in the Office of Endangered Species, had to stick with the job.

"I really shouldn't have because it was difficult to direct the office at the same time," he said. "But it's been fascinating, and I'm glad I did it." He added thoughtfully, "I want to save all life."

He said the Coachella Valley Fringe-toed Lizard Habitat Conservation Plan makes the second case in the nation of a conservation plan developed according to the conditions for a permit in the ESA. Definitely, the federal government has closely watched this story as a precedent for the next.

The story further shows government agencies that they too can work together. Leslie Cone, Indio area manager for the BLM and who has worked for the BLM for twelve years, said, "This is the finest example of cooperation of government on all levels I've ever been involved with." She feels man's needs come before trying to preserve every species on earth, but that it is "important to save representative examples of habitat—central parks for future generations to see without cement and asphalt."

Paul Wilhelm believes destiny preserved his oasis. Why else did Justin Dart hold the land for ten years? Why else did Nature Conservancy come to the valley right when the Lizard Club had begun to talk? Why was the spring day clear when the Mellon trustees explored Thousand Palms Oasis?

He said Indians still go to the oasis for healing ceremonies. "The Indians are very polite. They used to camp on the dunes so they wouldn't frighten animals away from the water." Maybe he thinks the old people, their spirits, have protected the oasis so that the Indians and animals may always come, and the desert may heal.

Al Muth would rather take destiny into his own hands. He said his fight for the fringe-toed lizard has been "a fascinating role. I guess you can now say...'How do you keep them down on the farm?' I cut my teeth on this project, gained insight into how and who one influences to get things done." As this story ended, he was thinking of new involvements. He mentioned nuclear waste. I asked if he'd like to join the battle against radioactive dumping. He said, "I don't know if I'd like to, or if I *have* to."

Somewhat ironically, the man who started the battle for a lizard habitat, Dr. Wilbur Mayhew, rarely visits the valley anymore. He trusts Muth to manage the research center and watchdog the developers. In charge of the University of California Natural Reserve System, Dr. Mayhew has been working to obtain seven new reserves. In his office the phone rang. He discussed fencing prices for one reserve. He hung up, and said to me, "I always planned to do that incubation work on desert reptiles—how moisture affects their reproduction—but never have. Too busy with habitat acquisition." Now, if he wanted to transport the protected Coachella Valley fringe-toed lizard to his laboratory, he'd have to seek a government permit.

In the secured realm of the fringe-toed lizard, the Coachella Valley Preserve, I reclaim a sense of untamed space I knew while growing up in the Coachella Valley. Bare mountains enfold the expanse of desert—a wilderness preserved.

IX.

EPILOGUE: A MODEL FOR THE FUTURE

"This is *déjà vu*," said Steve McCormick to the auditorium crowd. "Two-and-a-half years ago, in this very room, we were launching the project to solve the fringe-toed lizard crisis."

He joked about the lizard some people had called the "goddamn lizard." "At one point, we were scared appropriations from Congress would go to the 'french toast' lizard."

McCormick spoke to an audience invited by Nature Conservancy, the Palm Springs Desert Museum, the US-FWS, and the BLM to celebrate the creation of the Coachella Valley Preserve with awards and speeches, feasting and drinking.

He reminded those gathered that the odds had weighed against any fair resolution to the battle for lizard habitat. He said that when Nature Conservancy directors initially had analyzed how to acquire a desert preserve among golf resorts, they weren't encouraged.

"We were looking at taking on a $25 million project far outstripping anything we'd done before and which the BLM had said it couldn't do," McCormick told the crowd. "All interests were hostile, and the community was irate. Yet, an advisor who's worked a lot with Nature Conservancy said, 'If you're realistic, you'll never get anywhere with that project.'"

The audience chuckled, knowing that Nature Conservancy had instead dared to imagine a wetlands among sand dunes, and camaradarie among angry desert residents.

McCormick said that when he first met the embittered and diverse members of the Lizard Club nearly three summers ago, "I was almost uplifted with the realization that these people, coming from different backgrounds and protecting their own interests, together were the most creative, capable group I have seen."

That creative and unlikely group—ranchers and lawyers, biologists, bureaucrats, and developers—listened with growing euphoria because, from a complicated, nasty enmity, together they had created both a startling precedent of cooperation and also a rugged, spacious desert preserve.

McCormick paused. "It's a relief to have the project completed."

On that day, April 25, the Department of the Interior granted the ESA 10(a) permit allowing development on the critical habitat of the Coachella Valley fringe-toed lizard in exchange for a habitat preserve for the lizard. An official presented Riverside County Supervisor Corky Larson hefty copies to hand out to the nine cities and the county.

Having said goodbye to the rainy redwood country, Cameron Barrows arrived in the Coachella Valley early in 1986 to manage the Coachella Valley Preserve. His office is the Palm House at Thousand Palms Oasis. "After the isolation of the Nature Conservancy preserve I managed in the redwoods, I wasn't ready for an office in the middle of Palm Springs," he said.

Looking beyond his oasis headquarters, he admired "the long views in the desert, the soft colors on rocks, and the detail when you get up close to plants." He said half of him bemoans the development of the desert, while the other half rejoices because building brings mitigation fees to the preserve; the fees fund additional preserve acquisition and its management.

Within two months Barrows, along with formulating a management guide for the four owners of the preserve, arranged for two groups to clear the tamarisk chokehold away from oasis palm trees. Marina High School students, in a class entitled Crises in the Environment, "...got fed up with hearing about all the bad things people do to the environment, so the teacher told them some ways they could do something," Barrows said. Nature Conservancy calls the students, who came to the desert from the Southern California coast, the Crisis Team.

Barrows said the other group, the California Conservation Corp, "are from a lower-economic class. These are eighteen- to twenty-two-year-old, hard-line kids. I provide them with food, equipment. The state pays them."

Ric Wilhelm, the grandnephew of Louis Wilhelm and who helps manage the preserve, sauntered into the Palm House. "I've got the corp kids digging out tamarisk up the wash," he told Barrows. "I find I've got to really keep on them. I'll explain what to do and a few are joking around. Then I see them digging up what I said not to. I've got to stay right there. But I think they're going to work out." He nodded sagely. "Last night they went out for pizza. It'll be okay."

Instead of relying upon outside help, Barrows wants to build up a local volunteer program. When he meets people who've retired to the valley from other states, he tells them about Nature Conservancy preserves in their home states. He hopes they can transfer their appreciation for the preserved wilderness back home to the Coachella Valley Preserve, just as they have moved their furniture and hobbies. "I think women whose husbands now are playing golf all day will get involved with us," he said.

Amiable, retired visitors react differently to a lizard preserve from other people Barrows meets. While ninety percent of the preserve has been bought, mostly in big parcels from large land owners, small land owners hang on to the remaining lizard habitat. (See Map III) When Barrows arranged for fencing of the preserve, he asked these small owners for permission to fence out off-road vehicles from their property. "Some said it was okay. Most said, 'Forget it until you come up with a price I like.'" One man furiously responded he'd tell off-road vehicle users to rampage at will over the dunes and to kill any lizards they found.

Nevertheless, a good part of the preserve has been fenced. Impressive signs studded with the official seals of Nature Conservancy, the BLM, the USFWS, and the Department of the Interior, prohibit trespassing. Barrows himself hikes to the dunes and confronts violators. "A lot are kids who didn't know this was a special place, and once I explain that the desert is a harsh environment and that the noise and destruction from off-road vehicles makes animals expend a lot more energy just to survive, they go elsewhere. Others say, 'Sure, sure.' I know when I've turned my back, they'll be out there again."

Help will arrive when the CDFG and the USFWS send wardons to patrol the preserve. Barrows calls the preserve "a unique area in the nation because state and federal agencies, and private interests own and manage the preserve. Nature Conservancy will be doing more of this cooperation work with government as more of the western wild lands get into government hands. Nature Conservancy needs to develop relations with government now in order to see to it special lands are protected."

At the museum party to celebrate the difficult and extraordinary birth of the Coachella Valley Preserve, Steve McCormick said this preserve marks the "beginning of cooperative management. We've got a technique, an approach never before tried in this magnitude. Everyone's willing to jump in if everyone else chips in equally.

"In some ways it's thrilling, and in some ways disappointing that it's all over."

He said disappointing because the steering committee enjoyed developing trust and influence together, and hates to see a good job end. But this kind of working camaraderie will save wilderness again. A steering committee patterned after the diverse Coachella Valley Preserve model has already been negotiating to preserve the huge Carrizo Plains in Central California.

Not surprisingly, UC Riverside Professor Wilbur Mayhew has thrown himself into the core of the work towards preserving the Carrizo Plains. No doubt Dr. Mayhew, when the committee has nearly established the preserve, and as it carries out the final stages, surely will ignite his next drive to save a species and its habitat, just as thirty years ago he fired off the salvo that led to the battle over the Coachella Valley fringe-toed lizard. Only this time, he will have a model for success to go by.

110

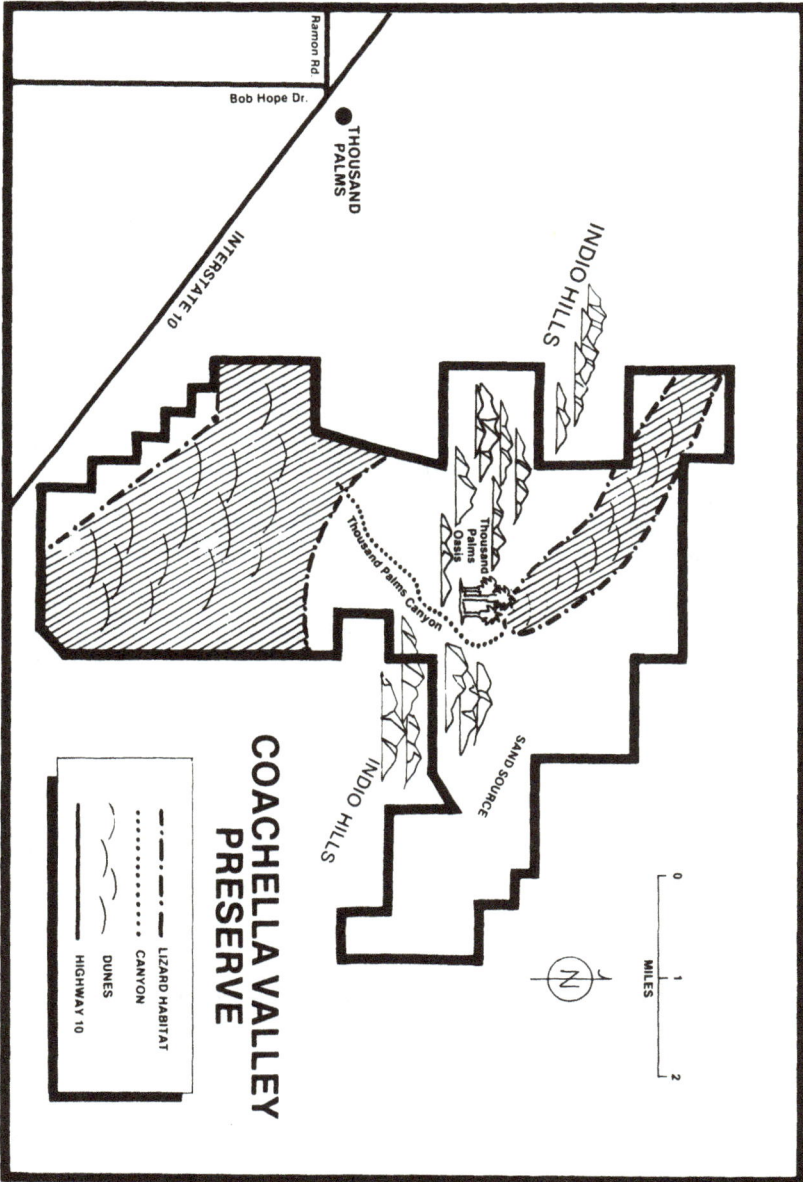

Map III: Coachella Valley Preserve

NOTES

1Gen. 1:28

2Ken Wells, "How a Little Lizard Gets Palm Springs to Do Its Bidding," *Wall Street Journal*, July 9, 1984.

3Coachella Valley Fringe-toed Lizard Habitat Conservation Plan Steering Committee, chaired by Nature Conservancy, *Coachella Valley Fringe-toed Lizard Habitat Conservation Plan*, June 1985, I-3.

4U.S. Fish and Wildlife Service, prepared by Thomas Reid Associates (Palo Alto, California), *Adoption and Implementation of Coachella Valley Fringe-toed Lizard Habitat Conservation Plan and Endangered Species Act Section 10(a) Permit Final Environmental Impact Statement*, February 1986, 17.

5Katherine Ainsworth, *The McCallum Saga/The Story of the Founding of Palm Springs* (Palm Springs: Palm Springs Desert Museum, 1973), 125.

6James W. Cornett, "Coachella Valley's Thousand Palms," *The Nature Conservancy News*, September/October 1984, 20.

7Jim Cornett, "Uma, the Fringe-toed Sand Lizard," *Pacific Discovery*, April/June 1983, 6.

8*Ibid.*

9*Ibid.*

10*Ibid.*

11Sid England, "Coachella Valley Fringe-toed Lizard Recovery Technical Review Draft," State of California Department of Fish and Game, presented by John M. Brody at the Fish and Wildlife Service public hearing July 7, 1980, Palm Springs, California, 1.

12Jim Cornett, "Uma, the Fringe-toed Sand Lizard," *Pacific Discovery*, April/June 1983, 5.

13Sid England, "Coachella Valley Fringe-toed Lizard Recovery Technical Review Draft," State of California Department of Fish and Game, presented by John M. Brody at the Fish and Wildlife Service public hearing July 7, 1980, Palm Springs, California, 1.

14Jim Cornett, "Uma, the Fringe-toed Sand Lizard," *Pacific Discovery*, April/June 1983, 4.

15*Ibid.*, 9.

16A. Sidney England and Steven G. Nelson, "Status of the Coachella Valley Fringe-toed Lizard (*Uma inornata*)," State of California Department of Fish and Game, Inland Fisheries Administrative Report No. 77-1, July 1976, 3.

[17]Sid England, "Coachella Valley Fringe-toed Lizard Recovery Technical Review Draft," State of California Department of Fish and Game, presented by John M. Brody at the Fish and Wildlife Service public hearing July 7, 1980, Palm Springs, California, 3.

[18]Peter Aleshire, "Lizards May Help Humans," *Desert Sun*, December 12, 1978, sec. A, p. 2.

[19]*Ibid.*

[20]*Ibid.*

[21]A. Sidney England and Steven G. Nelson, "Status of the Coachella Valley Fringe-toed Lizard (*Uma inornata*)," State of California Department of Fish and Game, Inland Fisheries Administrative Report No. 77-1, July 1976, 3.

[22]*Ibid.*, 7, 10.

[23]*Ibid.*, 24.

[24]*Ibid.*, 1, 24, 25.

[25]*Ibid.*, 25, 26.

[26]*Ibid.*, pp. 26, 27.

[27]*Ibid.*, 27.

[28]*Ibid.*

[29]*Ibid.*, 1, 27.

[30]*Ibid.*

[31]*Ibid.*, 27, 28.

[32]*Ibid.*, 1, 27.

[33]Steve Moore, "Builders Battle Anew with Friends of the Fringe-toed Lizard," *Press-Enterprise*, May 22, 1980.

[34]Leonard Metz, "Vanishing Lizard Imperils Housing Loans," *San Bernardino Sun Telegram*, October 28, 1978.

[35]"'Snow-shoe' Lizards Pose Loan Problem," *Indio Daily News*, October 28, 1978.

[36]Caleb Trainer, "Valley's Fringe-toed Lizard Draws Scientific Interests," *Desert Sun*, October 9, 1980.

[37]Tracy I. Storer, *General Zoology* (New York: McGraw-Hill Book Company, Inc., 1951), 256-257.

[38]*Ibid.*, 257.

[39]James W. Cornett, "Interbreeding between *Uma inornata* and *Uma notata*," *The Southwestern Naturalist*, May 1982, 223.

[40]Peter Aleshire, "Lizard "Listing' No Threat to Loans," *Desert Sun*, November 15, 1978, sec. A, p. 1.

[41]Chamber of Commerce of Indio, "Resolution of the Chamber of Commerce of Indio, California Opposing the Declaration of Large Portions of Coachella Valley as Critical Habitat for the Coachella Valley Fringe-toed Lizard," November 1, 1978, 2.

[42]"Lizard Still on Danger List Spokesman Claims," *Indio Daily News*, March 9, 1979.

[43]Martin Hill, "Lizard Reserve Is Endorsed," *Desert Sun*, March 28, 1979.

113

44Alice Z. Cuneo, "Thousands of Rare Species Lose Federal Protection," *Press-Enterprise*, March 7, 1979.

45Virgil Kraft, "Letter to the Editor: 'Lizards and Odd Fish,'" *Desert Sun*, July 7, 1980.

46"County Votes to Fight Lizard Habitat Proposal," *Indio Daily News*, November 29, 1978.

47"Realtors Seek Solution to Valley Lizard Crisis," *Indio Daily News*, December 12, 1978.

48Martin Hill, "Lizard Reserve Is Endorsed," *Desert Sun*, March 28, 1979.

49"2 Hearings Set on Fringe-toed Lizard Status," *Press-Enterprise*, June 15, 1980.

50Peter Aleshire, "Hearings Slated on Listing of Controversial Sand Lizard," *Indio Daily News*, June 16, 1980.

51Peter Aleshire, "State Says Lizard Is Endangered," *Indio Daily News*, June 28, 1980.

52Peter Aleshire, "Bank Loans Not Affected by Ruling," *Desert Sun*, November 17, 1978.

53Alex Yaron, "Lewis Urges Testimony on Lizard," *Indio Daily News*, July 2, 1980.

54Martin Salditch, "U.S. Designates Fringe-toed Lizard as Threatened Species," *Press-Enterprise*, September 20, 1980, sec. B, p. 10.

55*Endangered Species Act of 1973 as Amended by Public Law 97-304 (The Endangered Species Act Amendments of 1982)*, sec. 3(20).

56*Ibid.*, sec. 4(f).

57Wesley G. Hughes, "Lizard May Halt Desert Area Growth," *Los Angeles Times*, December 27, 1983, sec. 1, p. 3.

58*Endangered Species Act of 1973 as Amended by Public Law 97-304 (The Endangered Species Act Amendments of 1982)*, sec. 9(a)(B).

59*Ibid.*, sec. 3(19).

60Paul Wilhelm, "A Famous Desert Swap," *Desert Magazine*, September 1960, 15-18.

61Keith Carter, "Agreement Reached on Lizard Habitat," *Palm Desert Post*, September 25, 1980.

62*Endangered Species Act of 1973 as Amended by Public Law 97-304 (The Endangered Species Act Amendments of 1982)*, sec. 10 (B)(i).

63*Ibid.*, sec. 10(a).

64Glen Martin, "Reserved, Forever," *PSA Magazine*, September 1983.

65"Small Beginning Made on Fringe-toed Lizard Habitat," *Press-Enterprise*, March 26, 1981.

66"Firm Cited for Destroying Lizard Habitat," *Press-Enterprise*, March 31, 1984, sec. B, p. 2.

67"Congress Votes $4.9 Million for Preserve," *Desert Communities Newspapers, Inc.*, November 7, 1984.

68Steve Moore, "Ford Backs Fund-Raising for Desert Preserve," *Press-Enterprise*, May 9, 1984.

69"McCallum Foundation Grants $300,000 Toward Preserve," *Desert Sun*, November 16, 1984.

70Katherine Ainsworth, *The McCallum Sage/The Story of the Founding of Palm Springs* (Palm Springs: Palm Springs Desert Museum, 1973), 49-54.

71*Ibid.*, 110.

72*Ibid.*, 106.

73*Ibid.*, 223.

74*Ibid.*, 195-197.

75*Ibid.*, 209.

76*Ibid.*, 220.

77Steve Moore, "Conflict: Lizard vs. Investment," *Press-Enterprise*, August 9, 1983, sec. B, p. 1.

78"'Snow-shoe' Lizards Pose Loan Problem," *Indio Daily News*, October 28, 1978.

79Coachella Valley Fringe-toed Lizard Habitat Conservation Plan Steering Committee, chaired by Nature Conservancy, *Coachella Valley Fringe-toed Lizard Habitat Conservation Plan*, June 1985, S-1.

80Sam Ramirez, "C-VAG Panel Approves Lizard Protection Plan," *Desert Sun*, January 10, 1985.

81Sam Ramirez, "Association of Governments Calls for Hearings on Fringe-toed Lizard Area," *Desert Sun*, October 25, 1984, sec. A, p. 5.

82Sally Bowman, "Fringe-toed Lizard Plan Eludes C-VAG," *Desert Communities Newspapers, Inc.*, October 24, 1984.

83Sam Ramirez, "County Backs Action to Save Lizards," *Desert Sun*, January 4, 1985.

84Sam Ramirez, "Several Growers Resist Plan for Fringe-toed Lizard Preserve," *Desert Sun*, November 15, 1982, sec. A, pp. 1-2.

85Coachella Valley Fringe-toed Lizard Habitat Conservation Plan Steering Committee, chaired by Nature Conservancy, *Coachella Valley Fringe-toed Lizard Habitat Conservation Plan*, June, 1985, V-10.

86*Ibid.*, V-12.

87U.S. Fish and Wildlife Service, prepared by Thomas Reid Associates (Palo Alto, California). *Adoption and Implementation of Coachella Valley Fringe-toed Lizard Habitat Conservation Plan and Endangered Species Act Section 10(a) Permit Final Environmental Impact Statement*, February 1986, 76.

88Coachella Valley Fringe-toed Lizard Habitat Conservation Plan Steering Committee, chaired by Nature Conservancy, *Coachella Valley Fringe-toed Lizard Habitat Conservation Plan*, June, 1985, S-3.

89*Ibid.*, V-13.

90U.S. Fish and Wildlife Service, prepared by Thomas Reid Associates (Palo Alto, California), *Adoption and Implementation of Coachella Valley Fringe-toed Lizard Habitat Conservation Plan and Endangered Species Act Section 10(a) Permit Final Environmental Impact Statement*, February 1986, 17.

91Coachella Valley Fringe-toed Lizard Habitat Conservation Plan Steering Committee, chaired by Nature Conservancy, *Coachella Valley Fringe-toed Lizard Habitat Conservation Plan*, June 1985, V 12-15.

92U.S. Fish and Wildlife Service, prepared by Thomas Reid Associates (Palo Alto, California), *Adoption and Implementation of Coachella Valley Fringe-toed Lizard Habitat Conservation Plan and Endangered Species Act Section 10(a) Permit Final Environmental Impact Statement*, February 1986, 35.

93*Ibid.*, 42, 54, 55.

94Ken Wells, "How a Little Lizard Gets Palm Springs to Do Its Bidding," *Wall Street Journal*, July 9, 1984.

95"Paying a Call on the Coachella Valley Fringe-toed Lizard," *Sunset Magazine*, February 1985, 74-75.

96Mike Kataoka, "Planning Council OKs Edom Hill Water Park," *Press-Enterprise*, April 19, 1985, sec. B, pp. 1-2.

97Elenita Ravicz, "People to Swim in Section of Desert Where Lizards Now Scamper," *Press-Enterprise*, June 6, 1985, sec. B, p. 3.

BIBLIOGRAPHY

BY

M. LOUISE REYNNELLS
Rural Information Center, Beltsville, Maryland

"2 Hearings Set on Fringe-toed Lizard Status." *Riverside Press-En- terprise*, June 15, 1980.

Adoption and Implementation of Coachella Valley Fringe-toed Lizard Habitat Conservation Plan and Endangered Species Act Section 10 (a) Permit Final Environmental Impact Statement. Prepared for U.S. Fish and Wildlife Service by Thomas Reid Associates, Palo Alto, California, February 1986.

Ainsworth, Katharine. *The McCallum Saga: The Story of the Found- ing of Palm Springs.* Palm Springs, CA: Palm Springs Desert Mu- seum, 1973.

Aleshire, Peter. "Bank Loans Not Affected by Ruling." *Desert Sun*, November 17, 1978.

_____. "Hearings Slated on Listing of Controversial Sand Lizard." *Indio Daily News*, June 16, 1980.

_____. "Lizard 'Listing' No Threat to Loans." *Desert Sun*, November 15, 1978.

_____. "Lizards May Help Humans." *Desert Sun*, December 12, 1978, sec. A, p. 2.

_____. "State Says Lizard Is Endangered." *Indio Daily News*, June 28, 1980.

Amacher, G. S., Richard J. Brazee, J. W. Buckley, and R. A. Moll. *Application of Wetland Valuation Techniques: Examples from Great Lakes Coastal Wetlands.* Ann Arbor, MI: Institute of Water Research, Michigan State University, 1989.

_____. *An Interdisciplinary Approach to Valuation of Michigan Coastal Wetlands. Technical Report*, (G1429-02). Ann Arbor, MI: School of Natural Resources, Michigan State University, 1988.

Baltezore, James F., *et al. Status of Wetlands in North Dakota in 1990. Agricultural Economics Report*, (269). Fargo, ND: North Dakota Agricultural Experiment Station, Department of Agricultural Economics, North Dakota State University, 1991.

Bardecki, M. J. "The Economics of Wetland Drainage," in *Water Quality Bulletin*, Vol. 14(2), April 1989:76-80, 103-104.

117

Barrows, C. Monitoring Report: Coachella Valley fringe-toed lizard. Southern California area manager, The Nature Conservancy, 1989.

Beatley, Timothy. "Balancing Urban Development and Endangered Species: The Coachella Valley Habitat Conservation Plan," in *Environmental Management*, January/February 1992, v. 16 (1):7-19.

_____. "Land Development and Protection of Endangered Species: A Case Study of the Coachella Valley Habitat Conservation Plan. Prepared for the National Fish and Wildlife Foundation, February 1990.

Bell, F. W. *Application of Wetland Valuation Theory to Commercial and Recreational Fisheries in Florida.* Tallahassee, FL: Department of Economics, Florida State University-Tallahassee, 1989.

Bennett, J., and I. Goulter. "The Use of Multiobjective Analysis for Comparing and Evaluating Environmental and Economic Goals in Wetland Management," in *GeoJournal*, v. 18(2), March 1989:213-220.

Bergstrom, J. C., J. R. Stoll, J. P. Titre, and V. L. Wright. "Economic Value of Wetlands-Based Recreation," in *Ecological Economics*, v. 2(2), July 1990:129-147.

Bingham, Gail, *et al*, eds. *Issues in Wetlands Protection: Background Papers Prepared for the National Wetlands Policy Forum.* Washington, DC: World Wildlife Fund and The Conservation Foundation, 1990.

Blackburn, Wilbert H., and John G. King, eds. *Water Resource Challenges and Opportunities for the 21st Century: Proceedings of the First USDA Water Resource Research and Technology Transfer Workshop, August 26-30, 1991, Denver, Colorado.* Washington, DC: Agricultural Research Service, U.S. Department of Agriculture, 1991.

Borre, M. A. *Wetlands in the Lake Champlain Region of Vermont: Present and Future Threats to the Resource, Boundary Determination and Background Information for the EPA's Propsed Advanced Identification, Final Report.* New Haven, CT: School of Forestry and Environmental Studies, Yale University, 1988.

Bowman, Sally. "Fringe-toed Lizard Plan Eludes C-VAG." *Desert Communities Newspapers*, October 24, 1984.

Burke, David G., *et al.* *Protecting Non-Tidal Wetlands. Planning Advisory Service Report*, (412-13). Washington, DC: American Planning Association, 1988.

Cable, Ted T., Virgil Brack and Virgil R. Holmes. "Simplified Method for Wetland Habitat Assessment," in *Environmental Management*, v. 13(2), March-April 1989:207-213.

Carey, Marc, Ralph Heimlich and Richard Brazee. *A Permanent Wetland Reserve: Analysis of a New Approach to Wetland Protection. Agriculture Information Bulletin*, (610). Washington, DC: Economic Research Service, U.S. Department of Agriculture, 1990.

Carpenter, C. C. "Patterns of Behavior in the Forms of the Fringe-toed Lizards," in *Copeia* 2:406-412.

Carter, Keith. "Agreement Reached on Lizard Habitat." *Palm Desert Post*, September 25, 1980.

Chew, Matthew K. *Bank Balance: Managing Colorado's Riparian Areas. Bulletin of the Cooperative Extension Service of Colorado State University*, (553A). Fort Collins, CO: The Denver Service Center, 1991.

Coachella Valley Association of Governments. *Coachella Valley Area Growth Monitor*. Palm Desert, CA: CVAG, 1989.

Coachella Valley Fringe-toed Lizard Habitat Conservation Plan Steering Committee, chaired by the Nature Conservancy. *Coachella Valley Fringe-toed Lizard Habitat Conservation Plan.* June 1985.

"Congress Votes $4.9 Million for Preserve." *Desert Communities Newspapers*, November 7, 1984.

Conservation Directory: A List of Organizations, Agencies, and Officials Concerned with Natural Resource Use and Management. Washington, DC: National Wildlife Federation, 1991.

Cooper, David J. "Colorado's Wetlands," in *The Green Thumb*, v. 45(2), Autumn-Winter 1988:38-45.

Cornett, James W. "Coachella Valley's Thousand Palms." *The Nature Conservancy News*, September/October 1984:18-21.

_____. "Interbreeding between *Uma inornata* and *Uma notata*." *The Southwestern Naturalist*, May 1982:223.

_____. "Uma, the Fringe-toed Sand Lizard." *Pacific Discovery*, April/June 1983:3-10.

Costanza, R., S. C. Farber, and J. Maxwell. "Valuation and Management of Wetland Ecosystems," in *Ecological Economics*, v. 1(4), Dececember 1989:335-361.

"County Votes to Fight Lizard Habitat Proposal." *Indio Daily News*, November 29, 1978.

Cubbage, Frederick W., L. Kathrine Kirkman and Lindsay R. Boring. "Federal Legislation and Wetlands Protection in Georgia: Legal Foundations," in *Forest and Ecology Management*, v. 33(34), June 1, 1990:271-286.

Cuneo, Alize Z. "Thousands of Rare Species Lose Federal Protection." *Riverside Press-Enterprise*, March 7, 1979.

"CVWD to Add 1200 Acres to Preserve." *Riverside Press-Enterprise*, May 25, 1984.

Davis, Ronald B., and Dennis S. Anderson. *The Eccentric Bogs of Maine: A Rare Wetland Type in the United States*. Orono, ME: Maine Agricultural Experiment Station; Orono, ME: Department of Plant Biology and Pathology, Institute for Quaternary Studies, University of Maine-Orono, 1991.

Yvonne P. Tevis

Dahl, Thomas E. *Wetlands Losses in the United States, 1780's to 1980's.* Washington, DC: Fish and Wildlife Service, U.S. Department of the Interior, 1990.

_____, **and Craig E. Johnson.** *Wetlands Status and Trends in the Conterminous United States, Mid-1970's to Mid-1980's: First Update of the National Wetlands Status Report.* Washington, DC: Fish and Wildlife Service, U.S. Department of the Interior, 1991.

Domestic Policy Council's Task Force Report on Wetlands: Summary of Public Meetings and Written Comments. Washington, DC: Office of the Secretary, Department of the Interior, 1991.

Douglas, Aaron J. "Annotated Bibliography of Economic Literature on Wetlands," in *Biological Report,* (89 19). Fort Collins, CO: National Ecology Research Center, Fish and Wildlife Service, U.S. Department of the Interior, September 1989.

Dumont, Paul G. "Grasslands, Wetlands, and More Wildlife," in *Soil and Water Conservation News,* v. 12(2), July-August 1991:6-11.

Endangered Species Act of 1973 as Amended by Public Law 97-304 (The Endangered Species Act Amendments of 1982). Washington, DC: U.S. Government Printing Office, 1983.

England, Sidney. "Coachella Valley Fringe-toed Lizard Recovery Technical Review Draft." State of California Department of Fish and Game. Paper presented by John M. Brody at the Fish and Wildlife Service public hearing July 7, 1980, Palm Springs, California.

England, A. Sidney, and Steven G. Nelson. "Status of the Coachella Valley Fringe-toed Lizard (*Uma inornata*)." State of California Department of Fish and Game, Inland Fisheries Administrative Report (77-1), July 1976.

Farber, Stephen. "The Value of Coastal Wetlands for Recreation: An Application of Travel Cost and Contingent Valuation Methodologies," in *Journal of Environmental Management,* v. 26(4), June 1988:299-308.

Feierabend, S., and J. Zelazny. *Status Report on Our Nation's Wetlands.* Washington, DC: National Wildlife Federation, 1988.

Field, Donald W. *Coastal Wetlands of the United States: An Accounting of a Valuable National Resource.* Washington, DC: National Oceanic and Atmospheric Administration, U.S. Department of Commerce; Fish and Wildlife Service, U.S. Department of the Interior, 1991.

Finlayson, Max, and Michael Moser, eds. *Wetlands.* New York: Facts on File, 1991.

"Firm Cited for Destroying Lizard Habitat." *Riverside Press-Enterprise,* Mar. 31, 1984, sec. B:2.

Fisk, David W., ed. *Proceedings of the Symposium on Wetlands: Concerns and Successes, September 17-22, 1989.* Technical Publications Series, (89-3). Bethesda, MD: American Water Resources Association, 1989.

Folke, Carl. "The Societal Value of Wetland Life-Support," in *Linking the Natural Environment and the Economy: Essays from the Eco-Eco Group.* Carl Folke and Thomas Kaberger, eds. Boston, MA: Kluwer Academic Publishers, 1991, 141-171.

Frayer, W. E., D. D. Peters and H. R. Pywell. *Wetlands of the California Central Valley: Status and Trends 1939 to Mid-1980's.* Portland, OR: Fish and Wildlife Service, U.S. Department of the Interior, 1989.

Gnam, Rosemarie S. "Wetland Bills Swamp Congress," in *BioScience*, v. 4(3), March 1992:222.

Goldstein, Jon H. "The Impact of Federal Programs and Subsidies on Wetlands," in *Transactions of the Fifty-third American Wildlife Natural Resources Conference: New Approaches in Managing Natural Resources, Louisville, Kentucky, March 18-23, 1988.* Washington, DC: Wildlife Management Institute, 1988, 436-443.

Grafton, William N., et al. *Conference Proceedings: Income Opportunities for the Private Landowner Through Management of Natural Resources and Recreational Access.* Morgantown, WV: Extension Service, West Virginia University, 1990.

Gregory, Stan, and Linda Ashkenas. *Riparian Management Guide: Willamette Naitonal Forest.* Portland, OR: Forest Service, U.S. Department of Agriculture, 1990.

Gresswell, Robert E., Bruce A. Barton and Jeffrey L. Kershner, eds. *Practical Approaches to Riparian Resource Management: An Educational Workshop, May 8-11, 1989, Billings, Montana.* Billings, MT: Bureau of Land Management, U.S. Department of the Interior, 1989.

"Hearings Set on Fringe-toed Lizard Status." *Riverside Press-Enterprise*, June 15, 1980.

Heimlich, Ralph E. *National Policy of 'No Net Loss' of Wetlands: What Do Agricultural Economists Have to Contribute? Staff Report*, (9149). Washington, DC: Resources and Technology Division, Economic Research Service, U.S. Department of Agriculture, 1991.

_____. "Wetlands and Agriculture: New Relationships," in *Forum for Applied Research and Public Policy*, Spring 1991:78-83.

_____. "A Wetlands Reserve: What Cost," in *Agricultural Outlook-AO*, (166), August 1990:23-25.

Hey, D. L. "Wetlands: A Future Nonpoint Pollution Control Technology," in *Technical Publication*, (88-4). Bethesda, MD: American Water Resources Association, November 1988:225-235.

Hickman, Clifford A. "Forested-Wetland Trends in the United States: An Economic Perspective," in *Forest Ecology and Management*, v. 33(34), June 1, 1990:227-238.

Hill, Martin. "Lizard Reserve is Endorsed." *Desert Sun*, March 28, 1979.

Hollis, G. E., et al. "Wise Use of Wetlands," in *Nature and Resources*, v. 24(1), January-March 1988:2-13.

Hook, Donal D., et al., eds. *The Ecology and Management of Wetlands, vol. 1, Ecology of Wetlands; vol 2, Management, Use and Value of Wetlands.* Portland, OR: Timber Press, 1988.

————, **and Russ Lea, eds.** *Proceedings of the Symposium: The Forested Wetlands of the Southern University States, Orland, Florida, July 12-14. General Technical Report SE, (50).* Asheville, NC: Southeastern Forest Experiment Station, Forest Service, U.S. Department of Agriculture, 1989.

Hubbard, Daniel E. *Glaciated Prairie Wetland Functions and Values: A Synthesis of the Literature. Biological Report, (88 43).* Washington, DC: Fish and Wildlife Service, U.S. Department of the Interior, 1988.

Hughes, Wesley G. "Lizard May Halt Desert Area Growth." *Los Angeles Times*, December 27, 1983, sec. 1, 3.

Indio Chamber of Commerce. "Resolution of the Chamber of Commerce of Indio Opposing the Declaration of Large Portions of Coachella Valley as Critical Habitat for the Coachella Valley Fringe-toed Lizard." CCI, November 1, 1978.

Infoterra/USA: Directory of Enviromental Sources. Washington, DC: Information Sharing Branch, Information Management Division, Office of Resources Management, U.S. Environmental Protection Agency, 1991.

"Interim Habitat Conservation Plan for the Stephens' Kangaroo Rat." Prepared by the Regional Environmental Consultants, Riverside County Planning Department, 1990.

Jatnieks-Straumanis, Sarma A., and Lawrence E. Foote. "Wetland Mitigation Banking—How It Works in Minnesota," in *Rangelands*, v. 10(3), June 1988:120-123.

Kataoka, Mike. "Planning Council OKs Edom Hill Water Park." *Riverside Press-Enterprise*, April 19, 1985, sec. B:1-2.

Kauffeld, Judy. "A New Swamp Dawning," in *Ohio 21*, v. 5(1), March 1991:6-9.

Kellert, S. *Public Attitudes Toward Critical Wildlife and Natural Habitat Issues.* Washington, DC: U.S. Fish and Wildlife Service, 1979.

Kirali, Sari J., Ford A. Cross and John D. Buffington. *Federal Coastal Wetland Mapping Programs: A Report by the National Ocean Pollution Policy Board's Habitat Loss and Modification Working Group. Biological Report, (90 18).* Washington, DC: Fish and Wildlife Service, U.S. Department of the Interior, 1990.

Kraft, Virgil. "Letter to the Editor: 'Lizards and Odd Fish.'" *Desert Sun*, July 7, 1980.

Kusler, Jon. *Our National Wetlands Heritage: A Protection Guidebook.* Washington, DC: Environmental Law Institute, 1983.

_____. "Wetlands Delineation: An Issue of Science or Politics?" in *Environment*, v. 34(2), March 1992:6-11.

_____, **and Mary Kentula.** *Wetland Creation and Restoration: The Status of the Science.* Washington, DC: Island Press, 1990.

Kuzilla, Mark S., Donald C. Rundquist, and Jeffery A. Green. "Methods for Estimating Wetlands Loss: The Rainbasin Region of Nebraska, 1927-1981," in *Journal of Soil & Water Conservation*, v. 46(6), November-December 1991:441-446.

Leitch, Jay A., and Brenda L. Ekstrom. *Wetlands Economics and Assessment: An Annotated Bibliography.* New York: Garland, 1989.

"Lizard Still on Danger List, Spokesman Claims." *Indio Daily News*, March 9, 1979.

Lugo, Ariel E., Mark Brinson and Sandra Brown, eds. *Forested Wetlands (Ecosystems of the World, 15).* Amsterdam: Elsevier, 1990.

Lupi, Frank, Theodore Graham-Tomasi, and Steven J. Taff. *An Hedonic Approach to Urban Wetland Valuation. Staff Paper,* (91-8). St. Paul, MN: Department of Agricultural and Applied Economics, University of Minnesota, 1991.

_____. "Forested Wetlands in Freshwater and Saltwater Environments," in *Liminology and Oceanography*, v. 33(4), July 1988:894-909.

Lyke, William L., and Thomas J. Hoban, eds. *Proceedings of the Symposium on Coastal Water Resources.* Technical Publications Series, (88-1). Wilmington, NC: American Water Resources Association, 1988.

Majumdar, Shyamal K., et al. *Wetlands Ecology and Conservation: Emphasis in Pennsylvania.* Easton, PA: Pennsylvania Academy of Science, 1989.

Manci, Karen M. *Riparian Ecosystem Creation and Restoration: A Literature Summary. Biological Report,* (89 20). Washington, DC: Research and Development, Fish and Wildlife Service, U.S. Department of the Interior, 1989.

Marsh, Dale. "A Wetland...Is a Wetland...Is a Wetland," in *Journal of Soil and Water Conservation*, v. 43(4), July-August 1988:282-285.

Marsh, L. L., and R. D. Thornton. "San Bruno Mountain Habitat Conservation Plan," in *Managing Land Use Conflicts: Case Studies in Special Area Management*, D. J. Brower and D. S. Carol, eds. Durham, NC: Duke University Press, 1987.

Martin, Glen. "Reserved, Forever." *PSA Magazine*, September 1983.

"McCallum Foundation Grants $300,000 Toward Preserve." *Desert Sun*, November 16, 1984, sec. A:2.

123

Metz, Leonard. "Vanishing Lizard Imperils Housing Loans." *San Bernardino Sun Telegram,* October 28, 1978.

Miller, Brian K. *Wetlands, Regulations, and You: What Every Indiana Farmer Needs to Know. Final Report,* (128). West Lafayette, IN: Cooperative Extension Service, Purdue University, Jan. 1991.

Minnesota Wetland Evaluation Methodology for the North Central United States, Final Report. St. Paul, MN: Army Corps of Engineers, U.S. Department of Defense, 1988.

Moore, Steve. "Builders Battle Anew with Friends of the Fringe-toed Lizard." *Riverside Press-Enterprise,* May 22, 1980.

_____. "Conflict: Lizard vs. Investment." *Riverside Press-Enterprise,* August 9, 1983, sec. B:1.

_____. "Ford Backs Fund-Raising for Desert Preserve." *Riverside Press-Enterprise,* May 9, 1984.

_____. "Fringe-toed Lizard Flares Anew in Desert Area," in *Riverside Press-Enterprise,* July 2, 1983.

Motzkin, Glenn. *Atlantic White Cedar Wetlands of Massachusetts.* Research Bulletin Number, (731). Amherst, MA: Massachusetts Agricultural Experiment Station, College of Food and Natural Resources, University of Massachusetts, 1991.

Munro, J. W. "Wetland Restoration in the Mitigation Context," in *Restoration Management Notes,* v. 9(2), Winter 1991:80-86.

Muth, A. "Population Biology of the CVFTL." *Progress Report* (3), submitted to California Department of Fish and Game, July 1, 1989.

Myers, Lewis H. *Riparian Area Management: Inventory and Monitoring of Riparian Areas. Technical Reference,* (1737-5). Denver, CO: Denver Service Center, Bureau of Land Management, U.S. Department of the Interior, 1990.

New Mexico Riparian-Wetland 2000: A Management Strategy. Santa Fe, NM: New Mexico State Office, Bureau of Land Management, U.S. Department of the Interior, 1990.

Newton, R. B. *Forested Wetlands of the Northwest.* Environmental Institute Publication, (88-1), Amherst, MA: University of Massachusetts-Amherst, 1988.

Nicholas, Sara. "The War Over Wetlands," in *Issues in Science and Technology,* v. 8(4), Summer 1992:35-41.

Norris, P. E., S. Tully, and M. R. Dicks. *Wetlands Determinations, Definitions, and Management Options. OSU Current Report,* (876). Stillwater, OK: Cooperative Extension Service, Oklahoma State University, March 1991.

Olson, R. K., and K. Marshall, eds. *Workshop Proceedings: The Role of Created and Natural Wetlands in Controlling Nonpoint Source Pollution, Arlington, Virginia, June 10-11, 1991.* Corvallis, OR: ManTech Environmental Technology, 1991.

124

Parks, Tony R. "Wetlands Regulation and Timber Harvesting in the 1990's," in *Proceedings of the Nineteenth Annual Hardwood Symposium of the Hardwood Research Council: Facing Uncertain Futures and Changing Rules in the 1990's, Starkville, Mississippi, March 10-12, 1991.* Memphis, TN: Hardwood Research Council, 1991, 47-49.

"Paying a Call on the Coachella Valley Fringe-toed Lizard." *Sunset Magazine*, February 1985:74-75.

Pettry, David E. "Wetland Delineation: The Environment Problems and Solutions," in *Proceedings of the 19th Annual Hardwood Symposium....op cit.*, 51-56.

Protecting America's Wetlands: An Action Agenda. Washington, DC: The Conservation Foundation, 1988.

Ramirez, Sam. "Association of Governments Calls for Hearings on Fringe-toed Lizard Area." *Desert Sun*, October 25, 1984, sec. A:5.

_____. "County Backs Actions to Save Lizards." *Desert Sun*, January 4, 1985.

_____. "C-VAG Panel Approves Lizard Protection Plan." *Desert Sun*, January 10, 1985.

_____. "Several Growers Resist Plan for Fringe-Toed Lizard Preserve." *Desert Sun*, November 15, 1982, sec. A:1-2.

"Rancho Mirage Favors Desert Preserve." *Riverside Press-Enterprise*, October 3, 1983.

"Rancho Mirage to Sue Over Lizard." *Riverside Press-Enterprise*, November 18, 1983.

Ravicz, Elenita. "People to Swim in Section of Desert Where Lizards Now Scamper." *Riverside Press-Enterprise*, June 6, 1985, sec. B:3.

"Realtors Seek Solution to Valley Lizard Crisis." *Indio Daily News*, December 12, 1978.

Recovery Plan for the Coachella Valley Fringe-toed Lizard. Portland, OR: U.S. Fish and Wildlife Service, 1984.

Reed, Robert M., Martha S. Salk, and J. Warren Webb. *Evaluation of Impacts on Wetlands: Do NEPA [National Environmental Policy Act] Analyses Integrate Wetland Protection Requirements?* Oak Ridge, TN: Environmental Sciences Division, Oak Ridge National Laboratory, 1991.

Reeder, B., and W. J. Mitsch. *What is a Great Lakes Coastal Wetland Worth?: A Bibliography.* Technical Series. Department of Medicine, Assiut University Hospital, 1990.

Reid, T. S., and D. D. Murphy. "The Endangered Mission Blue Butterfly," in *The Management of Viable Populations: Theory, Applications and Case Studies*, B. Wilcox, P. Brussard, and B. Marcot eds. Stanford, CA: Stanford University, Center for Conservation Biology, 1986.

Riley, C. "Wetlands Preservation Policy Gains New Stature," in *Farmline*, v. 11(2), February 1990:4-7.

Riparian Management: *A Leadership Challenge.* Washington, DC: Forest Service, U.S. Department of Agriculture, 1991.

Riparian-Wetland Initiative for the 1990's. Washington, DC: Bureau of Land Management, U.S. Department of the Interior, 1990.

Salditch, Martin. "U.S. Designates Fringe-toed Lizard as Threatened Species." *Riverside Press-Enterprise,* September 20, 1980, B:10.

Salveson, David. *Wetlands: Mitigating, Regulating Development Impact.* Washington, DC: Urban Land Institute, 1990.

Sampson, R. Neil, and Dwight Hair, eds. *Natural Resources for the 21st Century.* Washington, DC: Island Press, 1990.

Saving Endangered Species: Implementation of the Endangered Species Act. Washington, DC: Defenders of Wildlife, 1987.

Schlaepfer, G. "A Model for Land Conservation: An Oral History Analysis of Habitat Preservation for the Coachella Valley Fringe-toed Lizard." M.A. thesis. California State University, Fullerton, California.

Scodari, Paul F. *Wetlands Protection: The Role of Economics.* Washington, DC: Environmental Law Institute, 1990.

The Second RCA Appraisal: Soil, Water, and Related Resources on Nonfederal Land in the United States—Analysis of Conditions and Trends. Washington, DC: Forest Service, U.S. Department of Agriculture, 1989.

Selzer, Paul T. "Habitat Conservation Plannning and the Coachella Valley Fringe-toed Lizard," in *Endangered Species Update,* 1989, v. 6, no. 10 (August):28.

Siegel, William C., and Terry K. Haines. "State Wetland Protection and Legilsation Affecting Forestry in the Northeastern United States," in *Forest Ecology and Management,* v. 33(34), June 1, 1990:239-252.

"Small Beginnings Made on Fringe-toed Lizard Habitat." *Riverside Press-Enterprise,* March 26, 1981.

"'Snow-shoe' Lizards Pose Loan Problems." *Indio Daily News,* October 28, 1978.

Soule, A. *How Effective Are Federal Programs in Mitigating Wetland Losses? Technical Report.* Tucson, AZ: Department of Hydrology and Water Resources, Arizona University, 1988.

Stark, Lloyd R. *A Study of Natural Wetlands Associated with Acid Mine Drainage. AML Research Contract Report, Pennsylvania State University,* (967409). Washington, DC: Bureau of Mines, U.S. Department of the Interior, 1990.

Stavins, Robert N. *The Welfare Economics of Alternative Renewable Resource Strategies: Forested Wetlands and Agricultural Production.* New York: Garland, 1990.

_____, and Adam B. Jaffe. "Unintended Impacts of Public Investments on Private Decisions: The Depletion of Forested Wetlands," in *American Economic Review*, v. 80(3), June 1990:337-352.

Stockdale, Erik C. *Freshwater Wetlands, Urban Storm Water, and Nonpoint Pollution Control.* Seattle, WA: Resource Planning Section, King County Department of Parks, Planning, and Resources, 1991.

Storer, Tracy I. *General Zoology*, second edition. New York: McGraw-Hill, 1951.

Strong, Mark A., J. Gregory Mensik and Daniel S. Walsworth. "Converting Rice Fields to Natural Wetlands in the Sacramento Valley of California," in *Transactions of the Western Section of the Wildlife Society: Papers Presented at the Joint Annual Meeting of the Western Section and the Northeast Section of the Wildlife Society*, v. 26, Eric R. Loft and Richard L. Callas, eds. Sacramento, CA: The Wildlife Society, 1990, 29-35.

Taff, Steven J. *What is a Wetland Worth?: Concepts and Issues in Economic Valuation. Staff Paper*, (92-1). St. Paul, MN: Department of Agriculture and Applied Economics, University of Minnesota, 1992.

_____, **and S. Todd Lee.** *Minnesota's RIM Reserve: Easement Summary and Payment Procedures. Economic Report*, (90-6). St. Paul, MN: Department of Agricultural and Applied Economics, University of Minnesota, 1990.

Tansey, John B., and Noel D. Cost. "Estimating the Forested-Wetland Resource in the Southeastern United States," in *Forest Ecology and Management*, v. 33(34), June 1, 1990:193-213.

Thorson, R. M., and S. L. Harris. "How 'Natural' Are Inland Wetlands?: An Example from the Trail Wood Audubon Santuary in Connecticut, USA," in *Environmental Management*, v. 15(5), 1991:675-688.

Tiner, Ralph W. *Wetlands of Rhode Island.* Newton Corner, MA: Fish and Wildlife Service, U.S. Department of the Interior, 1989.

Trainer, Caleb. "Valley's Fringe-toed Lizard Draws Scientific Interests." *Desert Sun,* October 9, 1980.

Tripp, J. T. B., and M. Herz. "Wetland Preservation and Restoration: Changing Federal Priorities," in *Virginia Journal of Natural Resources Law*, v. 7(2), Spring 1988:221-275.

Turner, Kerry, and Tom Jones, eds. *Wetlands: Market Intervention Failures.* London: Earthscan, 1991.

Turner, R. Eugene. "Landscape Development and Coastal Wetland Losses in the Northern Gulf of Mexico," in *American Biologist*, v. 30(1), 1990:89-105.

The Value of Wetlands: A Guide for Citizens. Chesapeake, VA: The Southeastern Virginia Planning District Commission, 1988.

van der Valk, Arnold, ed. *Northern Praire Wetlands.* Ames, IA: Iowa State University Press, 1989.

van Hees; Willem W. S. "Boreal Forested Wetlands—What and Where in Alaska," in *Forest Ecology and Management*, v. 33(34), June 1, 1990:425-438.

Wakely, J. S. *Mitigation Database: Tracking Mitigation Activities in the Section 404 Permitting Program, Final Report.* Vicksburg, MS: Army Engineer Waterways Experiment Station, Department of Defense, 1989.

Want, William L. *Law of Wetland Regulation.* New York: Clark Boardman Callaghan, 1989.

Warner Wetlands Area of Critical Environmental Concern (ACEC) Management Plan. Lakeview, OR: Lakeview District Office, Bureau of Land Management, U.S. Department of the Interior, April 1990.

Webster, R. W. "Habitat Conservation Plans Under the Endangered Species Act," in *San Diego Law Review*, 24:243-271.

Wells, Ken. "How a Little Lizard Gets Palm Springs to Do its Bidding." *Wall Street Journal*, July 9, 1984.

Wernstedt, K. *Wetlands and Wastewater Management: Questions, Answers, Advice and Guidance. Technical Report.* Washington, DC: Office of Cooperative Environmental Management, 1988.

Wetlands Legislation and Management, February 1988-October 1989. Springfield, VA: National Technical Information Service, 1988.

Wetlands: Meeting the President's Challenge, 1990 Wetlands Action Plan. Washington, DC: Fish and Wildlife Service, U.S. Department of the Interior, 1990.

Wetlands Protection: Bibliographic Series. Washington, DC: Information Management Services Division, Environmental Protection Agency, 1988.

Whigham, D. F., R. E. Good, and J. Kvet, eds. *Wetland Ecology and Management: Case Studies.* Boston, MA: Kluwer Academic Publishers, 1990.

Whitehead, J. C. "Measuring Willingness-to-Pay for Wetlands Preservation with the Contingent Valuation Method," in *Wetlands*, v. 10(2), 1990:187-202.

Wilcox, Douglas A., ed. *Interdisciplinary Approaches to Freshwater Wetlands Research.* East Lansing, MI: Michigan State University Press, 1988.

Wilen, Bill O., and Warren E. Frayer. "Status and Trends of U.S. Wetlands and Deepwater Habitats," in *Forest Ecology and Management*, v. 33(34), June 1, 1990:181-192.

Wilhelm, Paul. "A Famous Desert Swap." *Desert Magazine*, September 1960:15-18.

Williams, Michael. *Wetlands: A Theatened Landscape.* Cambridge, MA: Blackwell, 1991.

Yaffee, S. L. *Prohibitive Policy: Implementing the Federal Endangered Species Act.* Cambridge, MA: MIT Press, 1986.

Yaron, Alex. "Lewis Urges Testimony on Lizard." *Indio Daily News*, July 2, 1980.

Zelazny, John, and J. Scott Feierabend, eds. *Proceedings of a Conference: Increasing Our Wetland Resources, Mayflower Hotel, Washington, DC, October 4-7, 1987.* Washington, DC: National Wildlife Federation, 1988.

Zimmerman, John L. *Cheyenne Bottoms: Wetlands in Jeopardy.* Lawrence, KS: University Press of Kansas, 1990.

INDEX

Philip L. Boyd Deep Canyon Research Center, 9, 15, 27, 44, 56-57, 60, 102

pineal gland research, 35

Press Enterprise (Riverside), 48

protected species, 15, 59

R.A. Glass Company, 15, 73-74

Rancho Mirage, California, 17, 45, 48, 52, 90, 104

rare species, 16, 18, 37, 39, 41

recovery plan, 16, 51

Reid, Thomas—SEE: Thomas Reid Associates

Richard King Mellon Foundation, 16, 67-75, 77, 80, 88, 105

Riverside, California, 24, 56, 90

Riverside County Board of Supervisors, 8, 11, 29, 47, 51, 80, 103, 108

Riverside County Planning Department, 66

Sacramento Endangered Species Office—SEE: United States: Office of Endangered Species

Sahara Desert, 23-24

Salton Sea, 28

San Bernardino, California, 9, 53

San Gorgonio mountains, 27, 52, 55

San Jacinto mountains, 55

sand source, 28, 33, 45, 53, 98

Seligman, Pete, 68-69

Selzer, Paul, 9, 58, 61-63, 65-67, 70, 75, 77, 80, 92, 101-102

Sinatra, Frank, 26

Smoke Tree resort, 86

Southern Pacific railroad, 83

species, 42-43

"Status of the Coachella Valley Fringe-toed Lizard" (Sid England and Steven Nelson), 8, 38-41, 48

Stebbins, R. C., 34-35

Sunrise Company, 7, 9, 20, 58-62, 65-67, 69-71, 74-75, 77, 89

Sunset magazine, 102

"take," 16, 52, 59, 62, 64, 73-74

tamarisk, 34, 108-109

"third eye" research, 34-35

Thomas Reid Associates, 90

Thousand Palms, California, 42, 50-51

Thousand Palms Canyon, 28, 53, 87

Thousand Palms Chamber of Commerce, 9, 92

Thousand Palms Oasis, 9, 54, 68-70, 72, 79, 85-88, 91, 105-106, 108

threatened species, 16, 18, 37, 40-41, 51

Tierra Madre Consultants, 90

tumbleweed, 34

UCR—SEE: University of California, Riverside

Uma Inornata—SEE: Coachella Valley fringe-toed lizard

United States Fish and Wildlife Service, 7-8, 16, 18, 37, 41-42, 45-47, 50-52, 59, 62-64, 70, 73-74, 77, 90-92, 94, 97-99, 102, 107, 109-110

United States Bureau of Land Management, 7, 11, 27, 50-51, 55, 59, 62, 65, 70-71, 91, 95-96, 105, 107, 109

United States Department of the Interior, 108-109

United States House of Representatives, 8, 50-51, 98

ABOUT YVONNE P. TEVIS

YVONNE PACHECO TEVIS, a descendant of one of California's first governors, Romualdo Pacheco, developed a deep appreciation for the desert while growing up in the Coachella Valley community of Rancho Mirage. Both her father, a biologist, and mother were actively involved in desert politics, her father having served on the Rancho Mirage City Council and her mother having authored a how-to book on election campaigns at the local level. Fluent in Spanish, Tevis earned a bachelor's degree in English from the University of Redlands in 1983 and a master's degree in International Policy Studies from the Monterey Institute of International Studies in 1988. She served as a senior writer and editor for the University of California Institute for Mexico and the United States (UC MEXUS). This is her first book for The Borgo Press. She currently resides in San Francisco, California.

www.ingramcontent.com/pod-product-compliance
Lightning Source LLC
Chambersburg PA
CBHW031519270326
41930CB00006B/438

9 780893 704322